Scientific Forecasting and Human Needs

TRENDS, METHODS AND MESSAGE

Related Titles of Interest

DANZIN
Science and the Second Renaissance of Europe

FRIEDRICHS & SCHAFF
Microelectronics and Society: For Better or For Worse

GOLDSMITH & KING
Issues of Development: Towards a New Role for Science and Technology

GVISHIANI
Science, Technology and Global Problems

KING
The State of the Planet

LASZLO
Systems Science and World Order: Selected Studies

MAYOR
Scientific Research and Social Goals

FORTI & BISOGNO
Research and Human Needs

Scientific Forecasting and Human Needs

TRENDS, METHODS AND MESSAGE

Proceedings of a symposium held in Tbilisi, USSR
6-11 December 1981

EDITED BY
AUGUSTO FORTI

UNESCO, PARIS
AND
PERGAMON PRESS
OXFORD · NEW YORK · TORONTO · SYDNEY · PARIS · FRANKFURT

U.K.	Pergamon Press Ltd., Headington Hill Hall, Oxford OX3 0BW, England
U.S.A.	Pergamon Press Inc., Maxwell House, Fairview Park, Elmsford, New York 10523, U.S.A.
CANADA	Pergamon Press Canada Ltd., Suite 104, 150 Consumers Road, Willowdale, Ontario M2J 1P9, Canada
AUSTRALIA	Pergamon Press (Aust.) Pty. Ltd., P.O. Box 544, Potts Point, N.S.W. 2011, Australia
FRANCE	Pergamon Press SARL, 24 rue des Ecoles, 75240 Paris, Cedex 05, France
FEDERAL REPUBLIC OF GERMANY	Pergamon Press GmbH, Hammerweg 6, D-6242 Kronberg-Taunus, Federal Republic of Germany

First published 1984 by the United Nations Educational, Scientific and Cultural Organization, 7 Place de Fontenoy 75700 Paris, France, and Pergamon Press Ltd.

UNESCO ISBN 92 3 102134 6

First edition 1984

Library of Congress Cataloging in Publication Data

Main entry under title:
Scientific forecasting and human needs.
Essays presented at a symposium held in Tbilisi, USSR as part of the Unesco programme on research and human needs.
1. Science—Methodology—Congresses. 2. Forecasting—Congresses. 3. Technological forecasting—Congresses. 4. Social service—Congresses. I. Unesco. II. Title: Human needs.
Q174.S336 1984 303.4'9 83-25046

ISBN 0-08-027970-8

Printed in Great Britain by A. Wheaton & Co. Ltd., Exeter

Foreword

J. M. Gvishiani

The growing role of science and technology in all spheres of the material and spiritual life of modern society has become a characteristic feature of the present, dynamic time. A qualitatively new level of material and technical basis of production, brought about by scientific and technological breakthroughs, allows us to refer to it as an epoch of scientific and technological revolution. As a unit of revolutionary change in science and technology and an important generator of social changes, the scientific and technological progress directly and indirectly affects the position of man in production and non-production spheres, the structure of consumption and value judgements of individuals and social groups, and of peoples and civilizations. Naturally, in different social systems the scientific and technological change has different social promises and consequences, but in any case it considerably affects every modern society.

Reflecting the needs of social development, science not only accomplishes one of the indigenous human needs—the need for cognition of the surrounding world—but also actively participates in their formation and satisfaction. The determining role in this dialectical interrelationship is played by social and cultural factors. It is their influence that causes the seemingly paradoxical situation in which on the global scale the unprecedented achievements of science and technology are used rather ineffectively for the solution of really acute problems, and that the vital needs of a considerable part of mankind are not met so far, yet at the same time in certain social environments the scientific and technological progress is accompanied by the emergence of pseudo-needs, consumerism, and a rapacious attitude towards nature. A significant amount of material and human resources, scarce energy and raw materials is spent to meet these pseudo-needs which estrange the individual from society and do not contribute to his creative self-realization and spiritual development.

The acceleration of scientific and technological progress and its growing impact on man result in a greater emphasis being placed on the problems of humanistic orientation of science. Humanization of science and its closer orientation towards real, vital needs of individuals and society at large, have become one of the indispensable prerequisites for the utilization of unprecedented technological opportunities with a view to improving the living conditions of human beings, solution of crucial global problems and prevention of an impending nuclear holocaust. Both the orientation of science and application of its achievements are primarily set by social and political factors: as long as there are influential forces interested in directing a considerable share of scientific and technological potential of mankind, financial means alloted to science, and the skilled scientists needed to develop more and more powerful means of mass destruction, there cannot be any reliable guarantees of scientific humanization. Hence, the struggle for humanization of science is inseparably bound up with the struggle for detente, for lasting peace, mutual understanding and cooperation between all the peoples. Only in these conditions can there be sound opportunities for uniting the efforts of the scientific community, and of all the progressive forces of the globe towards the solution of the pressing current problems.

The stepping-up of the arms race, and the development of more sophisticated means of mass destruction, directly affects modern science and scientists, and leads to growing militarization. This means that more and more scientists, and an ever-growing share of resources intended for science, will not serve the purpose of meeting the immediate material and spiritual needs of humanity but will be directed towards the creation of means for their destruction. It is hardly possible to imagine anything more contradictory to the interests of human beings than this monstrous "anti-need".

Hence, today, as never before, special responsibility rests with scientists themselves, for a mobilization of their efforts towards the prevention of an arms race and towards the humanistic orientation of science. This is the major duty of the world scientific community.

A high level of science development and the expansion of its capacities in coping with vital current problems give grounds for belief that, under favourable circumstances and in conditions of effective cooperation, science will live up to the expectations relative to the human needs problem.

A special role in the present environment is played by international organizations which, irrespective of the changes in world policy, contribute to the creation and strengthening of a solid foundation for constructive, lasting cooperation of scientists from all over the world. Without doubt, among the organizations Unesco and IIASA are distinguished by their broad, interdisciplinary perception of current problems and the mobilization of scientific effort on their solution.

One of the characteristics of the present day is the aggravation of a number of problems which have acquired a truly international, global nature. It is worth pointing out that the solution of all these problems is so dependent on the ability of humanity to reasonably employ its powerful scientific and technological potential that the very problem of utilizing science and technology for the purposes of social development can justly be considered one of the crucial global problems of modern times.

Accordingly, it is quite natural that there is a growing concern of the public at large over the perspective evaluation of scientific and technological progress from the point of view of their contribution to the solution of current problems. Today, general changes, not only in technical and economic areas but also in social and psychological spheres of society, and the ways of its development, have become considerable factors determining our future. Nowadays, when the present immediately becomes the past, and when there are swift changes in our daily lives, only permanent orientation towards the future will allow us to meet the requirements of science.

Speaking of the role of science in the future, we certainly must be ready to face new, possibly more complex, problems. This is the logic of development, and it urges us to formulate scientifically based techniques for forecasting scientific and technological progress and its social implications.

The knowledge of objective trends in scientific cognition, based on a thorough consideration of its internal logic and laws, is a prerequisite not only for elaboration of a feasible national scientific policy but also for the allocation of material and human resources, as well as—and this is especially significant—for the timely training of scientific personnel. It also provides opportunities for a better perception of a future society where science will not only be a factor of socioeconomic development but also an important element in modern man's culture. This implies a growing significance of the social management of science, and humanistic utilization of its achievements for the benefit of man, for the satisfaction of his vital needs, provision of appropriate living conditions for creative work, and the self-realization necessary for man to consider himself worth of society and an active participant in the process of the societal development. At the same time, an artificial stimulation of material human needs, accompanied by suppression and sometimes ignorance of his spiritual needs, is the reason for blaming scientific and technological progress, and not its social implementation, for the alienation of man taking place in some industrially developed countries.

The aforementioned, and other, negative phenomena accompanying the growing scale of the application of scientific and technological achievements are often used as pretexts for the cultivation of a negative attitude towards science *per se*, for ostensibly depriving the human culture of its spiritualism, bringing about still new problems and not contributing to solution of the old ones. In fact with modern, highly institutionalized science, planned and managed at the national level, it is not really possible to blame scholars for the negative consequences of scientific and technological progress nor, moreover, to place upon them full responsibility for the insufficiently humanistic orientation of science. Nevertheless, the desire to exonerate science in the face of such accusations is undoubtedly one of the aspects characterizing the modern development of mass consciousness of scientists.

The growing social significance of science and its findings has resulted in the fundamental reflex reactions within the world scientific community which are accomplished by gradual formation of a new vision of science in modern society and its responsibility before it. This is manifested by a more extensive participation of the scientific community in the anti-war and progressive social movements. It is also reflected in the more militant stance of scientists towards modern crucial sociopolitical problems, in understanding the social interests they have in common with other groups of working people. Finally, it is reflected in the content of scientific activities themselves, the selection of socially significant topics of research, and a more responsible attitude towards how the results of the research will be used.

One of manifestations of the greater social responsibility is the ever-closer attention paid by scientists representing different disciplines and schools, to the evaluation of scientific and technological development with regard to its impact on man and society.

As was previously mentioned, at a time when the present quickly becomes the past, only a permanent orientation towards the future allows us to meet present day requirements. The swift pace of scientific and technological development, and the resulting socioeconomic changes, have turned scientific and technological forecasting from semi-fantastic guesses to an independent branch of research having an ever-greater impact on specific social and political decisions.

This practical orientation sets more stringent requirements on the effectiveness and validity of the methods of scientific and technological forecasting and their applicability to real-life decision-making. At the same time, such orientation raises the social responsibility of those contributing to their development.

An ever-greater comprehension of the growing social responsibility, on the part of scholars, for the direction of scientific and technological development, and for a maximum contribution of science to the satisfaction of material and spiritual needs of man, was testified by the International Seminar "Scientific Forecasting and Human Needs: trends, methods and message", held under the auspices of Unesco on 6-11 December 1981 in Tbilisi, the capital of the Georgian Soviet Socialist Republic. Along with the USSR State Committee for Science and Technology, the USSR Academy of Sciences, the All-Union Research Institute for Systems Studies, the USSR Commission for Unesco and a number of other Soviet organizations, a considerable contribution to the Seminar was made by the International Institute for Applied Systems Analysis (IIASA), IFIAS and representatives of some other international organizations. Despite a relatively small number of participants, this ensured rather broad representation and provided a basis for pointed and fruitful discussions.

The Seminar covered a wide range of problems connected with scientific and technological forecasting, dynamics of human needs as such, and their mutual influence. Realizing the impossibility of covering all the aspects of this *complexe problématique* the Organizing Committee decided to focus primarily on the methodological problems of forecasting and research into the human needs as well as evaluation—on the basis of specific national and international projects—of the results and perspectives of development and the application of different methodological approaches. There was also a discussion on some key processes, with respect to their implications for the individual and society, in the current development of science and technology linked with robotization and the utilization of microprocessors in industry. It was around these three areas that the presentations and discussions were centred.

Forecasting two very dynamic and complex processes such as scientific and technological development and the trends in the formation and satisfaction of the material and spiritual needs of man, especially given their complex and conflicting interrelationships, is far from easy. At present there are far more unresolved problems than ready recommendations and answers. Hence it becomes very important to unite the efforts of scientists representing different countries and research institutions, different disciplines and branches of science. Only close international and interdisciplinary co-operation can enable science to accomplish the complex tasks it faces at the present point in history.

For all the differences in opinions on, and approaches to, solution of some or other research problems—and I believe it is these differences that determined the fruitfulness of the discussions—all the participants of the Tbilisi Seminar shared a sincere yearning to contribute to the solution of the crucial problems of today, the concern for the humanistic orientation of science, and the realization of their social responsibility to the contemporary and forthcoming generations. Participants were unanimous concerning the need for more extensive and long-term international cooperation in the study of the problems discussed, providing the opportunities not only for an exchange of experience and data, but also for joint research by multinational teams of researchers. A striking illustration of such an international interdisciplinary body able to investigate the complex problems inherent at the present time is IIASA, which in 1981 commemorated its tenth anniversary. The indispensable prerequisites for the further intensification of constructive cooperation among scientists are, naturally, lasting peace on the globe, and greater confidence and mutual understanding between individuals and peoples.

I would like to express my sincere gratitude to the hospitable hosts of the Seminar—the Government, Academy of Sciences and the State Committee for Science and Technology of the Georgian Soviet Socialist Republic. The fact that the Seminar, devoted to the humanistic orientation of science, was held on Georgian soil seems to me rather symbolic. Georgia is a Republic with ancient historic and cultural traditions imbued with high optimism, a love for life and a deep faith in the inexhaustible spiritual capacities of man. The prominent Georgian poets, artists, composers and sculptors are famous all over the world. Highly developed science and technology go hand in hand with an extremely solicitous attitude towards nature, and the historical and natural relics with which the Republic is richly so endowed. World recognition has been won by Georgian scientists in the fields of psychology, archaeology, physics, mathematics and other branches of science. A noticeable interest was shown by the participation of the representatives of the Georgian Academy of Sciences.

I would like, in the name of the Seminar participants and with all my heart, to wish the Georgian people further successes. I would also like to express my deep gratitude for the participation in the organization, preparation and conducting of the Seminar to the members of the International and Soviet Organizing Committees, to Mr A. M'Bow, Director-General of Unesco and Mr A. Kaddoura, his Scientific Assistant, to Dr Federico Mayor who, as the First Deputy Director-General of Unesco, contributed much to the successful organization of the Seminar and the attraction of prominent scholars to it. I would also like to thank Mr A. King, President of IFIAS, whose active participation in the Seminar as a Chairman and speaker was a considerable contribution towards its success, to Professor K. Holling, Director of IIASA, as well as to Dr A. Forti, the Coordinator of the Unesco Programme on "Scientific Research and Human Needs", to Professor S.P. Kapitsa, and to all the Seminar participants.

Academician J. Gvishiani

Preface

An international symposium on Long-range Scientific Forecasting and Human Needs was organized within the framework of Unesco's programme on Research and Human Needs and held in Tbilisi, USSR, on 6-11 December 1981. The main objective in doing so was to bring together those scientists who were involved in recent major national and international long-range scientific forecasting exercises (such as, for example, the EEC "FAST" Model, "Interfutures", "Global 2000 and the "USSR Comprehensive Programme of Scientific and Technological Progress and its Social Implications for a 20-year period"), in order to confront varying methodologies and discuss the rôle which Unesco might play in serving the international community in this new and crucial area.

Unesco is particularly grateful to the USSR Academy of Science and the USSR State Committee for Science and Technology, who hosted the meeting, and the Georgian Academy of Science who helped in making the Symposium the successful scientific event it was, assembling a large number (more than forty) outstanding scientists from all over the world.

The opinions expressed in this publication are those of the authors, and do not necessarily reflect the views of Unesco.

Contents

Introduction

A. Forti

"Scientific Forecasting and Human Needs"—why bring together these two branches of modern science which, during recent years, have already been developing their own separate approaches, methods and techniques, and make this two-fold theme the focus of an international symposium? The close interrelationship between science, its trends, accomplishments and defects, as a solution to an important global problem-satisfaction of human needs-is now evident to all. What is new here is the fact that the *tempo* of change in our ever-changing world is so rapid that timely forecasting in both spheres becomes essential and indispensable in terms of the contribution of science to the satisfaction of human needs. Moreover, in the absence of forecasting, there would be no relevant approach to the subject.

The necessity of scientific forecasting for better long-term orientation of the decision-making process presents its practitioners with the challenge of making forecasting "less of an art and more of a science". Approaches to forecasting future developments of science and their implications in relation to human needs range across a wide spectrum that includes very sophisticated modern techniques, formal and non-formal scenarios, as well as combinations of a variety of approaches. What is important is the ever-increasing general awareness that achievement of at least some desirable alternatives rests upon awareness, that action rests on decisions, that decisions rest on forecasting and that forecasting as such can be useful if it is based on valuable objective data and scientific approaches.

To make research efforts in scientific forecasting related to problems of human needs more relevant, it is absolutely necessary to be aware and to make explicit, from the very outset, to whom these results would be addressed in terms of potential for practical application.

First of all, scientific forecasting is very important for the educational system, due to the fact that this system is potentially inertial. Hence, in order to re-orient its structure and programmes with a view to making it more relevant to the problems that will face science in the future, it is imperative that those responsible for it be informed in time.

This is a crucial point because it is not only the structure of the educational system in terms of specialization and training in various scientific disciplines which is concerned. Orientation of programmes also strongly depends to a greater extent on what society expects and demands from science as a whole. For example, the very nature of contemporary problems-which are inherently complex and need both comprehensive vision and a broad interdisciplinary approach-makes it imperative to re-orient university programmes in the natural sciences, incorporating in them inputs from the social and human sciences as well as interdisciplinary methods such as systems approach.

The impact science can make upon the satisfaction of human needs and attainment of societal goals depends considerably on the structure of modalities of scientific personnel and higher education. Due to the ever-increasing role of science in the solution of social problems, it is important to forecast the demand for scientific personnel and skills in line with the problems that science will face tomorrow. All this poses serious problems to the current systems of higher education which are the source of potential scientists and scholars of tomorrow. Special emphasis should be placed on a wide interdisciplinary outlook on the part of specialists and a good command of systems techniques that will ensure adequate research into these problems.

Secondly, the results of scientific forecasting appear to be indispensable to policy-makers. This fact is illustrated, for instance, by the challenges and problems linked with the ever-increasing utilization of microprocessors techniques and genetic engineering. Knowledge of the outcome of scientific forecasting exercises is therefore equally important for politicians in office and those temporarily out of office, who, being less directly involved with everyday problems, could obtain extra advantages in comparison to their more preoccupied colleagues by acquiring the longer-term overview of future problems and alternatives arising from scientific forecasting. This competition would, probably, make society less shortsighted and more aware of the challenges imposed by the accelerating *tempo* of change caused by scientific-technological development.

Last but not least, the results of forecasting are of great importance for scientists and research workers themselves; especially young ones, whose social responsibility implies an awareness of the social relevance of science as a whole and, their own research activity in particular.

There is a consensus developing in the scientific community, and amongst decision-makers, that keener foresight and more responsible forecasting are potentially desirable alternatives that have been overlooked and some decisions being made today could have negative and perhaps disastrous consequences tomorrow. Moreover, they could be realized so quickly that people might find they are powerless to protect themselves. The scientific community responds to this challenge by inventing or rediscovering new techniques and approaches directed towards gaining a greater insight into the real opportunities for choice and into the implications of the use of these opportunities.

Research in the field of course involves making judgments—the process whereby values, attitudes and beliefs guiding our expectations and preferences must come into play. These judgments should be exercised from the very beginning in the identification of the factors relevant to particular problems, relationships and goals that should be considered or, alternatively, could be eliminated. The best solution to this situation, where subjectivity is inevitable, is to admit and make it explicit, and advise the decision-maker of it. It is especially important because the outcome of this kind of scientific exercise is not the proposal of decisions, but rather the presentation of possible alternatives, noting their respective advantages and flaws.

Concerning methodological problems, it is important to "sharpen" the question of human needs, concentrating research efforts on the most crucial aspects of the human needs dynamics, taking into account that there are many different kinds of "needs", both short- and long-term. In fact, they represent, a hierarchical system at the highest level of which there are needs connected with survival (food, shelter, reproduction), followed by wealth, freedom, justice, knowledge, human potential, etc. Even at the highest level, there are inconsistencies between short-term needs (food and shelter) and long-term (e.g. reproduction, survival of the race, etc.): for instance, population growth interferes with the ability to exist. Research on human needs must be concentrated on these inconsistencies and incompatibilities.

The most urgent of collective human needs, equally important for all peoples, is the need for peace and prevention of wars, since lasting peace is not only a prerequisite for survival of mankind and the full realization of human potentials, but also a necessary precondition for the mobilization of the scientific potential of the planet for the solution of the most crucial social problems, and for satisfaction of material and spiritual human needs. Militarist orientation of scientific development represents not only meaningless misuse of human creativity, but also one of the main obstacles to the harmonization of the interrelations between science and man. Hence, the struggle for peace and disarmament has become an important sphere of social responsibility of scientists in the contemporary world. A more active cooperation is needed in this important field, and a multidisciplinary approach and interdisciplinary methods are indispensable for a comprehensive vision of this complex subject.

With reference to the contributions science can provide for the satisfaction of vital human needs, one should be fully aware that science is not only a very powerful means of solving crucial problems of the contemporary world, but also represents itself in one of the most ancient human needs-the need to know and understand. And as soon as the needs related to survival have been more fully satisfied, the role of this spiritual need will increase, since it contributes considerably to the self-realization of individuals. In contrast to material needs, spiritual needs such as art, culture, and science, are inherently non-saturable and their development can contribute to mitigating the dangerous tendency towards unlimited and non-satisfying consumerism.

Nowadays, fundamental science contributes directly to satisfaction of human needs but cannot be evaluated only by this yield. At the same time, taking into account the impossibility of foreseeing not only the practical application of the research outcome but also the results themselves, as well as the fact that unpredictability was and is the driving force in the advancement of science, certain scientific efforts and expenditures should always be devoted to scientific research not directly linked to any immediate practical purpose.

Public attitudes towards science—its benefits and its dangers for man and society—are becoming more and more important in the context of the modern world. While stressing the importance of concentrating research efforts on the problems related to satisfaction of the most urgent human needs, one should at the same time avoid an overly utilitarian, instrumental attitude to science and its role in human society.

At the present time, hundreds of institutes are involved with projects in scientific forecasting, and the range of approaches is now very diffuse. Nevertheless, this area is mostly viewed as a branch or sphere of application of systems analysis. Representing one such method, systems approach provides modern scientists with an effective tool which, over recent years, has been more and more widely used, not only in scientific-technological forecasting, but also in the study of human and societal needs and their dynamics. The importance of further elaboration of systems methods and their application to the problems of scientific forecasting in relation to human needs have been underscored by several contributors to this volume. In this context, of special significance are the choice of adequate techniques, the correct combination of formal and non-formal, quantitative and qualitative methods, and substantial profound interpretation of results and data obtained.

Systems analysis itself is not very easy to define, since it is definitely not a particular activity, but rather an *approach* to a particular activity. But what is important here as far as scientific forecasting is concerned is the fact that it is policy-oriented and problem-oriented, hence, normative, in the sense that its application could be efficiently related to particular problems. And some of the most crucial problems nowadays are those linked with human needs.

The application of the systems approach, which is inherently future-oriented, is impossible without conscious use of models appropriate to particular problems, and taking into account not only valuable and reliable information, but also judgment and, to a lesser degree, intuition.

Since scientific forecasting oriented to any particular problem or scope of problems is essentially problem-oriented, it cannot be based on particular disciplines, but demands an interdisciplinary outlook and participation of teams of specialists from fields as diverse as engineering, economics, mathematics, social and political studies and philosophy.

Special emphasis is given to modelling as the most developed and widely accepted technique for forecasting socioeconomic processes on the national, regional and global scale. The construction of models, reflecting the properties of the system under study and its goals, is one of the traditional principles of systems approach. Models reflecting a certain holistic view of the object and relationships of its major sub-systems and components, are always based on a certain concept reflecting the world outlook of the researchers. All models are implicitly normative and conditioned by the socio-psychological milieu. Explicit acknowledgment of this, by those who develop and use models, is necessary to avoid misunderstanding.

A significant factor considerably affecting the quality of results is the level of model disaggregation, i.e. the degree of detailing in the process description. One would think that by using more information the results would thereby be more accurate. However, by increasing the number of parameters which describe the behaviour of a real-life system, and accordingly complicating the model, one risks crossing the so-called "interpretation threshold", i.e. losing the ability to adequately perceive and evaluate the results.

On the other hand, the utilization of highly aggregated models is fraught with serious errors of a different kind: subconscious or conscious "projection into the future" of those trends, structures, interrelations and ideas, needs and values that seem habitual in the present-day world. These "paradigms of the present", built in the highly aggregated models, inevitably distort the picture of the future, for development, as is known, is not confined to purely quantitative changes but implies a fundamental rearrangement of the structure itself and the interrelations between the individual socioeconomic processes. It is these processes that play the determining role here.

So far, the models have mainly concentrated on economic indicators as an objective or yardstick, and very little attention has been given to human and social indicators. In the course of the Symposium's discussions it also appeared that some considerations of overriding importance had been overlooked in models of the world's future developed over the last decade, such as political and social tensions or conflicts, the problems of disarmament and its influence on the solution of global problems, issues of global climatic variations, etc. Disarmament, as participants have observed would open new opportunities due to emergence of extra resources, employed presently for the arms race and for maintaining from the point of view of the need in resources and the possibilities of their release. On the other hand, curbing the arms race will obviously contribute to harnessing inflation and other problems.

By and large, broad discussion of the use of the forecasting techniques in general, and of modelling in particular, developed so far indicated that there are many methodological problems to overcome. At the same time, as emphasized in the discussions, these methods are essential not only for adequate investigation of contemporary problems but also in practical decision-making processes. Hence, further improvement of these techniques to make them more efficient for practical implementation is imperative.

The usefulness of systems approach and models in this field does not mean that other kinds of forecasting should be replaced. Moreover, it was unanimously stressed by the participants that the role of non-formal methods in general and scenarios in particular is ever-increasing, as many problems and processes under study are either inherently non-formalized or could not be formalized for practical reasons, e.g. because of a lack of relevant data. A noticeable interest in this connection represents the positive and negative experience gained in implementation of a number of international projects such as "interfutures" carried out by a group of scientists from OECD member states, and FAST (EEC). In particular, participants at the Symposium emphasized the organizational difficulties evolving from the activities of multinational research teams which reflected the contradictions and lack of agreed-upon positions of the governments concerned as to issues under study. At the same time, there was a consensus among participants that it is only through international cooperation that the efforts in scientific forecasting relating to human needs could be really efficient and useful.

The topics of the Symposium embraced various aspects of scientific-technological forecasting related to human needs which have been traditionally studied within the framework of different scientific disciplines and branches. It is this interdisciplinary nature of the topics that is a main source of methodological difficulties facing scientists dealing with these issues. The essence of this challenge is to elaborate and implement approaches which, being adequate for the complex, still rather ambiguous and inherently interdisciplinary problems, would simultaneously provide comprehensive vision based on scientifically valid information and methods without replacing them by pseudo-objective speculations and premature conclusions. These difficulties implicitly or explicitly influenced not only reports and the course of discussions, but also the scope of problems and variety of different approaches. This awareness of the complexity of the problems and of the necessity of making them explicit, rather than hiding them behind apparently scientific arguments, seems to be not only an important aspect of social responsibility of scientists dealing with qualitatively new phenomena, but also a very important prerequisite for success in studying them.

Fully aware of these methodological problems and difficulties, we decided not to impose on this book an artificial order which could give an unrealistic picture of "the state of the art" in this area. The volume contains, in addition to the papers, a Summary discussion, and the Final document, and is divided into three parts:

1 methodological problems of scientific forecasting related to human needs;

2 experiences gained in realization of certain projects dedicated to investigation of these problems;

3 some modern trends in scientific-technological progress and its social implications.

The Symposium brought together for the first time people from various institutions representing different disciplines and countries, schools and approaches, and it was this possibility to discuss and compare various methods, to assess the results from the perspective of their social and humanistic relevance, which seems to be the most important outcome of the Symposium. The participants unanimously emphasized the necessity of organizing a forum of scientists dealing with these problems on a permanent, continuous basis, stressing the special role of Unesco as both interdisciplinary and intercultural organization in coordinating activity in this field.

Part I
Methodological Problems

A basic needs approach to impact analysis and evaluation for long-term policy-making

H. Bossel, B. Hornung and K. F. Müller-Reissmann

THE NECESSITY OF QUALITATIVE SYSTEMS ANALYSIS FOR POLICY-MAKING AND PLANNING

The evolution of human societies depends on man's interactions with his natural and social environment. These interactions are determined partly by more or less immutable laws or quasi-laws (from the laws of physics to social rules and role expectations) and partly by man's own free choices. Policy analysis—as the analysis of possible human actions in this environment, which are shaping the future—must therefore be concerned with both. Unfortunately the success of the natural sciences in the past has led to a misplaced emphasis on the "law" paradigm in the behavioural sciences. The resulting over-emphasis on quantification and the search for quasi-natural laws of human behaviour in experimental settings and statistical data has produced information of only limited usefulness to policy analysis. This state of affairs will remain until the paradigm of human choice as a result of reflection and reasoning gains greater scientific attention.[1]

Cognitive processes and variables determining behaviour are basically qualitative. The same is true for the normative assessment of the impacts of behaviour and events. Therefore computer simulation of such processes which are essential parts of long-term policy-making and forecasting is quite difficult. This is one of the main reasons why conventional systems analysis neglects the cognitive and in particular the normative aspects of policy-making and socioeconomic processes.

Normative assessments are central components of policy-making, planning, and of social research dealing with problems of the development of human societies. Decisions, the choice of behavioural strategies, and the assessment of social and economic change as development all depend on normative criteria.

* The paper is based on research carried out with the generous support of the Volkswagen Foundation at the Institute for Applied Systems Research and Prognosis (ISP), Königstr. 50A, 3000 Hannover 1, Federal Republic of Germany.

Valid criteria for what is good or bad for a society cannot be found by uncritically accepting existing value systems or by empirical research in the form of some type of opinion poll. Such methods reflect the ethnocentric and subjective ideas of people, which often do not encompass the real problems of a society.[2]

Orientation theory[3] avoids these difficulties by combining theoretical reflections about biocybernetic systems with the empirical analysis of existing value systems, which embody the knowledge about functional requirements of human societies which has been elaborated in collective learning processes. In this framework "values" are not simply subjective desires but represent the objective functional conditions of survival, without denying that they are shaped by cultural, ideological, and biographical influences and circumstances.

Orientation theory suggests a set of abstract basic values (basic orientors) which cover the whole spectrum of values.[3] The basic orientors indicate, at an abstract level, which requirements have to be fulfilled in order to make long-term survival of a system in a changing environment possible and more likely. Depending on the system and the problem under investigation, the set of basic orientors can be specified and elaborated at more concrete levels so that it can be used directly to assess the impacts of policies in specific fields, e.g. energy and environmental policies.[4] In this way orientation theory can provide guidelines for the elaboration of an encompassing set of criteria (orientors) at different levels of abstraction which are not selected at random but in a systematic way.

The non-numerical programme DEDUC allows use of such an orientor hierarchy for computer-assisted evaluation and assessment of impacts, though it is in most cases not possible to formulate quantitative relationships. Investigating and planning the future with such an instrument is no guarantee for success, but the use of relatively crude DEDUC models and evaluations based on orientation theory will help to avoid gross mistakes with regard to the functional requirements of human and social systems. An example of an analysis with a DEDUC model referring to the problem of decentralization of the economy in a modern society will be presented in the last section of this paper.

DEDUC is based on the logical, deductive processing of verbal concepts. The advantage of using DEDUC is that it allows one to deal with very complex problems in which a large number of variables and relations has to be considered, but about which precise knowledge is not available. It can only provide a rough survey over a problem area, however, and in many cases detailed quantitative research cannot be replaced. Nevertheless, it is beyond any doubt a considerable advantage to be able to gain such a rough and encompassing overview by using the computer as an incorruptible medium. In this way one can avoid excluding from the analysis whole problem areas at the very outset of research, due to an accidental choice of data, lack of hard data, limitations of the human capacity of information processing, mechanisms of psychological repression, and other irrational influences.

POLICY-MAKING AND PLANNING AS DEDUCTIVE PROCESSING OF CONCEPTS

In order to be able to deal with policy-making problems a cognitive system needs to have the following most important functional components:

(1) <u>The "internal model" of reality</u>. The internal model constitutes a

cognitive, more or less adequate representation of reality in the form of objects, respectively classes of objects as well as their properties and the (causal) relations between them. The internal model enables an actor to explain factual situations and, what is especially important for policy-making, to make predictions. This means that it enables him to analyse a problem situation as well as single alternative courses of action with regard to their expected consequences (subsequent states of the system).

(2) The "normative system". This contains the criteria used to valuate the (perceived) current and/or future factual situations which can be anticipated by means of the internal model. These criteria are suitable cognitive representations of desired circumstances.

If the situation in which policy-making takes place does not indicate alternative courses of action itself, a

(3) Generator of possible alternative courses of action is also required. In the simplest case it may merely consist of a memory, from which the alternatives can be taken one after the other in order to check their suitability by means of the internal model and the normative system. In addition to a simple memory humans dispose of heuristic strategies which can be used to synthesize new *ad-hoc* alternatives.

The following paragraphs discuss the roles which the internal model and the normative system play in the valuation of alternative courses of action as well as in the analysis of problems, or situations. As the functioning of both processes is the same, only the valuation of alternative courses of action is dealt with. Formally the analysis of problems (or situations) may be conceived as the valuation of actions: It may be seen as a valuation of what will happen if the actor does not act. After all, not to do anything is one possible course of action.

The valuation of action consists of two different cognitive components which have to be distinguished:

(1) A prognostic component: anticipation of future states of reality by means of the internal model (investigation of consequences).

(2) A valuating component: present and/or anticipated future states of reality are projected onto the normative system (valuation of consequences).

The process of valuation corresponds to the model of "teleologic ethics", which permits the valuation of an action only on the basis of its (expected) consequences. The consequences are being judged according to more or less general value criteria (orientors). This is in contrast to "deontic ethics", according to which the valuation of actions depends exclusively on whether certain valid, more or less concrete and precise norms are being observed or not.

Both components of a teleologic decision—investigation of consequences and valuation of consequences—can be conceived as *deductive* processes. In order to investigate consequences, general factual knowledge, drawn from the internal model, is being processed with regard to the particular case under consideration. Formally, the projection of a factual consequence (or an event) onto a value (orientor) can also be represented in the form of an implication (event-value-implication). This is also true for the relations between lower and higher orientors in an orientor hierarchy (value-value-implications).

Consequently it is also possible to represent processes of valuation as deduction processes, which are accessible to computer simulation at a justifiable expense. For this purpose the program DEDUC has been developed.

THE INTERACTIVE PROGRAM DEDUC

DEDUC is an interactive program for the deductive processing of verbal concepts.[5] At the beginning of the dialogue it does not contain any information about the real world, but only procedures serving to receive such information, to store it in appropriate internal data structures, to process it logically-deductively, to deliver outputs, etc. With the exception of the processing of certainty factors, load factors, and the factors indicating the degree of affectedness all procedures are list-processing, non-numerical procedures. In interaction with the computer the user can feed into the program his knowledge about reality ("factual knowledge") or his concepts of any problem area he chooses. He can also use concepts concerning the problem areas of other actors. The user subsequently orders DEDUC to process such concepts, delete or correct them, print out stored information, etc. For these purposes the user disposes of a simple dialogue language.

The concepts which are usually presented in natural language have to be appropriately formalized for computer processing. The formalism chosen for DEDUC is based on first-order predicate calculus. In comparison with this, however, DEDUC is simpler in form and more restricted in its possibilities of expression. It has also to be emphasized that any formalization of sentences of a natural language implies an interpretation giving such sentences a definite and unambiguous meaning. The formalization implying such interpretation has to be done by the *user*.

The formalization of concepts chosen for DEDUC contains three different possibilities of representation or of forms of input into the system: object definitions, implications, and premises.

(1) Object definitions: are used for the input of classificatory concepts in the form of objects, or object classes and their subclasses and specimen. For any class of objects a number of subdivisions or subclasses can be specified. The subclasses do not have to be disjunct (humans, for example, can be subdivided into men, women, children, adults, etc.).
Example:
NRENRG(FSSENRG(COAL,OIL,GAS) ,URAN),
RGENRG(SUB,WIND,WATER,BIOM) *IS* PRMENRG;

"Non-regenerative energies and regenerative energies are subclasses of the object class 'primary energy'; fossil energies and uranium are subclasses (or specimens) of the object class 'non-regenerative energies'; coal, oil, and natural gas are subclasses (or specimens) of the object class 'fossil energies'; energies from sun, wind, water, and biomass are subclasses (or specimens) of the object-class 'regenerative energies'."

(2) Implications: are used for the input of "composed" concepts, which can be represented in form of IF-THEN relations. It is irrelevant whether such a relation is more causal, consecutive, conditional or only an

identification or definition.

Examples:

IF DCCENTRL(:ECONOMY,COUNTRY,TIMES)[6]
THEN DCTRAFIC(CAR,COUNTRY,+TIMES) 90;

"For all countries and all times it is true that a decrease of centralization of the economy results in a decrease of car traffic in the next time period."

IF DCTRAFIC(CAR,COUNTRY,PLTIME)
THEN DCCONSUM(OIL,COUNTRY,+PLTIME) 80;

"A decrease of car traffic results in a decreasing consumption of oil."

(3) Premises: are used for the input of simple (i.e. not logically composed) predicate statements. Premises serve to make simple statements which are taken to be true or false and which are used in order to describe scenario conditions.

Example:

PREM DCCENTRL(:ECONOMY,WGERMANY,PRESENT);
PREM ICPRICE(:ENRG,INDNAT,PLTIME);

"Decentralization of the economy in Western Germany in the present."

"Increase of prices for energy in the industrial nations during the planning time (present and near future)."

The program generates new concepts, so-called conclusions, by logically deducing from the three types of concepts above. The deduction starts with one or several premises and tries to arrive at new, simple statements which are as general as possible, using the implications. Syntactically, conclusions have the same structure as premises and are being used in the same way: after being generated each conclusion serves in turn as a premise for the deduction of further conclusions. The process of deduction takes place in two directions: at a horizontal level it follows the logical structure of the implications, and vertically it takes place along the relations between classes and subclasses as defined by the object structures. The latter process is a classical deduction in the sense of a transition from the general to the particular, whereas the former follows the classical rules of how to draw conclusions, the *"modus ponens"* and the *"modus tollens"*.

If we take the concepts listed above as examples, and if we define in addition PRESENT, "near future" (NFUT), and "middle future" (MFUT) as subclasses of TIMES, and WGERMANY (Federal Republic of Germany) as a specimen of INDNAT (industrial nations), and the latter in turn as subclass of COUNTRY, DEDUC generates the following conclusions:

DCTRAFIC(CAR,WGERMANY,NFUT)
DCCONSUM(OIL,WGERMANY,MFUT)

"Decreasing car traffic in Western Germany in the near future."

"Decreasing consumption of oil in Western Germany in the middle future."

In a knowledge module implications and premises can be furnished with a so-called certainty factor (CF). The certainty factor is a rough measure for the degree of subjective certainty with which the concept in question is held to be true. Certainty factors run between the numerical values of 0 and 100. A concept with the certainty factor of 100 is taken to be absolutely certain. The other extreme, 0, means that there are no indications whatsoever that the implication might be valid. In the deduction process such implications are treated as non-existent.

The internal processing of certainty factors takes place according to the following rules:

(1) CF(P AND Q) : = min (CF(P), CF(Q))
The certainty factor of a conjunction equals the minimum of the certainty factors of the AND-section of the statement.

(2) CF(P OR Q) : = max (CF(P), CF(Q))
The certainty factor of a disjunction equals the maximum of the OR-section of a statement.

(3) If we take "*IF* L *THEN* R" for an implication, then is true:
CF(R) : = CF(L) × CF(*IF* L *THEN* R)/100

The certainty factor of the right side of the implication (conclusion) results from the certainty factor of the left side of the implication multiplied by the certainty factor of the implication divided by 100.

(4) The deduction of a certainty factor via negation is not possible. (In this respect the logic of certainties differs from the logic of probabilities.)

This means that results of deduction in the knowledge module are conclusions with specific subjective certainties (consequences of situations or alternative courses of action which had been put into the system in the form of premises). In order to valuate the consequences, they can be transferred to an orientation module by means of an interface. In principle an orientation module also consists of object structures, implications, and conclusions transferred from the knowledge module. The conclusions are to be valuated and serve as premises for the deductions in the orientation module. Syntactically the orientors are being treated as objects which are interconnected by means of implications (value-value-implications). They constitute a more or less sophisticated orientor hierarchy. Furthermore, the orientation module contains implications which serve to project the factual consequences generated in the knowledge module onto the lowest level of the orientor hierarchy (event-value-implications).

Each implication in the orientation module is furnished with a load factor (LF) which runs between the numerical values of -100 and +100. This load factor is a very rough subjective measure for the "intensity" of the relation between an event and an orientor, or between orientors of lower levels and those of higher levels. The higher the numerical value of this factor, the more the orientor on the right side of the implication is satisfied (+) or, as the case may be, violated (-) if the orientor on the left side is satisfied or violated.

Each premise (or conclusion) in the orientation module has a factor indicating the degree of affectedness (AF) which is a very rough subjective measure for the "intensity" with which the orientor in question is affected (satisfied or violated). The load factors of the implications are used to calculate the degree of affectedness for the conclusions generated from these implications.

The most important rule for processing these factors is the following:

If we take for an implication "*IF* L *THEN* R", the degree of affectedness is:

$$AF(R) := AF(L) \times LF(\textit{IF}\ L\ \textit{THEN}\ R)/100$$

The degree of affectedness of the right side of an implication (conclusion) equals the degree of affectedness of the left side, multiplied by the load factor of the implication divided by 100. If more than one implication loads onto one and the same orientor, the respective degrees of affectedness are added at the lowest level of certainty involved. Positive and negative values of the degree of affectedness are added separately. The different certainty factors are results of the knowledge module and correspond to the certainty factors of the consequences which had been valuated. Within the orientation module these certainties are not being changed any more.

DEDUC offers the possibility of presenting the results of the valuation under different aspects in a standardized and clearly arranged way in order to be able to compare as easily as possible the results of different alternatives or to compare the results of one and the same alternative in different conceptual systems. The user may choose any one of the levels of the orientor hierarchy for his investigation. If he chooses the uppermost level, he gets a highly aggregated and very general valuation. If he chooses a lower level the picture becomes correspondingly more detailed and more concrete. For the orientors on the level chosen the user has to specify weights between -100 and +100. A first diagram of evaluation shows merely the strength of the impact (satisfaction or violation) the chosen alternative and its consequences have on the orientors under investigation ("top values"). A second diagram includes the weights of the orientors, and a third also includes the certainties with which the orientors are affected. In a fourth diagram all single impacts affecting an orientor with different certainty factors are being aggregated into one impact which represents the maximum product $AF \times CF$ which can be found among the single impacts (see Figure 3). Furthermore it is possible to obtain a valuation over all time periods, in which the result for the particular periods can be weighted differently (see Figures 1 and 2).

These different ways of presenting the results of the valuation make it easier to localize and identify the causes for the differences which appear if different conceptual systems are being used to valuate one and the same course of action. Are such differences mainly due to differences in the knowledge module, or are they due to differences between the certainties with which some of the facts are believed to be true; or is it that the orientors and their weights are mainly responsible, or maybe the relations between the orientors?

IMPACT EVALUATION OF A POLICY OF DECENTRALIZING THE ECONOMY OF THE FEDERAL REPUBLIC OF GERMANY

In this section some results of DEDUC runs are discussed which were made with a model constructed for research in the field of political decision-making in the Federal Republic of Germany.[7] This model reflects the ideal-typical cognitive system of the leaders of the conservative party (CDU/CSU) in the Federal Republic with regard to future developments in the fields of economic, energy, and environmental problems.

The model is based on the analysis of documents, programmes, publications, etc. of the respective party, as well as on questionnaires administered to a number of its active politicians. Corresponding models for two other political groups are also available. The conservative party is of particular interest, however, because its assessment of the consequences of a decentralization of the economy in the Federal Republic can be analysed in this model with regard to four time periods which are defined by the following sequence of a time chain:

```
PRIT;
THE TIME CHAIN IS:
 PRESENT (present)
 NFUT    (near future)
 MFUT    (middle future)
 DISTFUT (distant future)
```

An event, for example, which happens in the present has its immediate consequences in the near future. From those consequences result new effects in the middle future, etc. For the present study two scenarios were analysed.

Scenario 1 contains the assumption that efforts towards a decentralization of the economy are made in the present. In this case the input into the model is the premise:

```
PREM DCCENTRL(:ECONOMY,WGERMANY,PRESENT);
```

Scenario 2 is characterized by an additional assumption that during the present and near future the prices for energy will rise. Inputs into the model are in this case the following two premises:

```
PREM DCCENTRL(:ECONOMY,WGERMANY,PRESENT),
PREM ICPRICE(:ENRG,INDNAT,PLTIME);
```

In the knowledge module Scenario 1 produces 18 consequences and Scenario 2 results in 33 consequences. If we look at the consequences which a decentralization of the economy (Scenario 1) will have in the middle future according to this model we find:

```
PRIPSO MFUT;
  6) CF:100 DCCENTRL(ADMIN,WGERMANY,MFUT)
 12) CF: 72 DCCONSUM(OIL,WGERMANY,MFUT)
 13) CF: 59 ICINCOME(WGERMANY,MFUT)
 14) CF: 72 DCMONOT(WORK,WGERMANY,MFUT)
 15) CF: 90 ICAUTON(LOREUNIT,WGERMANY,MFUT)
OK;
```

During this period the consequences are throughout positive: decentralization of the administration, decreasing oil consumption, increasing income, decreasing monotony at work, and increasing autonomy of local and regional societal and economic units. In Scenario 2 the decreasing consumption of oil contributes to reducing the negative effects of the rising prices of energy.

Figures 1 and 2 show the evaluation of the consequences of the two scenarios in the orientation module. The figures show "eval 7", the most highly aggregated evaluation which takes into consideration the certainty factors, load factors, and the weights of the top level orientors (the weights used are given in the figures in the column "WT".[8]

Furthermore, the two figures show the sum of the evaluations of all time periods. For better comparability the different time periods are weighted equally in the present examples (all weights are 100). It is possible, however, and depending on the problem under investigation sometimes even indispensable, to give less weight to future time periods and, for example, give a weight of 90 to the near future, 50 to the middle future and only 10 to the distant future. "eval 7" keeps the positive and negative effects separately and does not provide their sum total.

```
EVAL 7;
 PRESENT      TIME WEIGHT:   100
 NFUT  TIME WEIGHT ?
100
 MFUT    TIME WEIGHT ?
100
 DISTFUT  TIME WEIGHT ?
100

 VALUE     |WT. |CF-|CF+ | AGGR. EVALUAT. OVER TIME (AGGEVT)
 _________|___|___|____|________________________________________________
          |   |   |    -250 -200       -100         0         100        200  250
 _________|___|___|____||____|__________|__________|__________|__________|____|
 HEALTH   | 75|  0|  54|                            0+++++++++
1 WEALTH  | 55|  0|  50|                            0++++
 SELFREAL| 80|  0|  68|                             0+++++++++++++
2 INTPREST| 40| 50|  75|                           -0+++++
 LIBSUBSY| 45|  0|  90|                             0++++++++++
3 LIBSOC  | 60|  0|  50|                            0+++
 ECOLEFF | 65|  0|  75|                             0+++++
 ECONEFF | 80| 50|   0|                          ---0
4 EFFCTRL | 45| 94|  85|                            0++++++++
 SAFEDAN | 60|  0|   0|                             0
5 SAFESUPL| 70|  0|  75|                            0+++++++++++++++
 FLXIB   | 75| 94|  80|                             0++++++++++++++++++++++444
6 CPINNOV | 75|  0|  78|                            0++++++++++++++++++++++
 SOCRESP | 60|  0|  75|                             0++++
7 RESPFUT | 80|  0|  75|                            0+++++++

    PRESENT   :100     NFUT:100     MFUT :100     DISTFUT:100
```

Figure 1. Scenario 1: decentralization of the economy. Evaluation of consequences in all time periods, with equal weight (100) given to each period.

Basic orientors represented by the top values are:
1. Physical needs
2. (Social-psychological) needs
3. Freedom of action
4. Efficiency
5. Security
6. Adaptivity
7. Ethical basic orientors

If we compare the evaluations of the two scenarios in Figures 1 and 2 we see that according to this model decentralization of the economy has positive effects on nearly all orientors (with the exception of "economic efficiency" and "safety from dangers", the latter not being affected at all). The relatively massive positive evaluation is the same in Figures 1 and 2. This means that the additional premise of rising energy prices, which is used in Scenario 2, has only negative effects. These show up clearly in the comparison of the two diagrams.

In Figure 1 only two orientors, "economic efficiency" and "international prestige", are affected negatively with -30 and -10 points. In Figure 2, on the other hand, 13 orientors are being affected negatively. But also in this scenario the positive effects are much stronger than the negative effects, both looking at the whole picture and looking at the individual orientors, with two exceptions. With regard to the fact that the negative consequences are largely due to rising energy prices this evaluation means that efforts towards reducing the centralization of the economy would be a strategy advantageous with regard to most orientors.

```
EVAL 7;
 PRESENT      TIME WEIGHT:  100
 NFUT  TIME WEIGHT ?
100 100 100
 MFUT   TIME WEIGHT ?
 DISTFUT  TIME WEIGHT ?

 VALUE     !WT. !CF-!CF+  ! AGGR. EVALUAT. OVER TIME (AGGEVT)
 _________!___!___!____!_____________________________________________________
          !    !    !   -250 -200       -100          0          100        200  250
_____________!___!___!____!!___!_________!__________!__________!__________!____!
 1   HEALTH   ! 95! 50!  54!                          ---0+++++++++
_____WEALTH   ! 55! 70!  51!                        -----0++++
 2   SELFREAL ! 80! 50!  69!                        ----0+++++++++++++
_____INTPREST ! 40! 74!  75!                          ---0+++++
 3   LIBSUBSY ! 45!  0!  90!                             0++++++++++
_____LIBSOC   ! 60! 50!  50!                           --0+++
     ECOLEFF  ! 65! 75!  75!                           --0+++++
 4   ECONEFF  ! 80! 81!   0!                 -----------0
_____EFFCTRL  ! 45! 94!  85!                             0++++++++
 5   SAFEDAN  ! 60!  0!   0!                             0
_____SAFESUPL ! 70! 50!  75!                            -0+++++++++++++++
 6   FLXIB    ! 75! 77!  80!                            -0++++++++++++++++++++++444
_____CPINNOV  ! 75! 75!  78!                            -0++++++++++++++++++++++
 7   SOCRESP  ! 60! 75!  75!                            -0++++
_____RESPFUT  ! 80! 75!  75!                           --0+++++++

   PRESENT   : 100    NFUT: 100    MFUT : 100    DISTFUT: 100
```

Figure 2. Scenario 2: decentralization of the economy and increasing energy prices. Evaluation of consequences in all time periods, with equal weight (100) given to each period.

Positive effects are concentrated in the basic orientor dimensions "(social-psychological) needs" (no. 2 in the diagrams) and "adaptivity" (no. 6). Also strongly positively affected is "safety of supplies".

The total evaluation over all time periods shown in Figure 2 can be decomposed into the separate evaluations of the four time periods used in the time chain. Two of them, the evaluations for the near and middle future, are shown in Figure 3. Figures 2 and 3 show that most of the positive effects of Scenario 2 appear in the near future, whereas the majority of negative effects appears in the middle future. On the negative side only "economic efficiency" is strongly affected in the near future.

The study of the effects of decentralizing the West German economy shows that even in a model reflecting the cognitive system of a relatively sceptical group of policy-makers decentralization has considerable positive effects on a set of orientors derived from orientation theory. This means that it would be a strategy apt to advance most of the basic needs of the society while negative impacts are quite limited.

The application of DEDUC as presented in this chapter demonstrates how two different policies or scenarios can be compared with regard to their long-term impacts on all parts of a society. It also shows how the impacts of different factors, whether they are "given" scenario conditions, real conditions or policy measures, can be isolated. The comparison of Figures 1 and 2 allows one to distinguish the impacts of rising energy prices from those of a decentralization of the economy itself. In the present case, where Scenario 1 consists only of one premise and Scenario 2 of the same plus one additional premise, this may seem relatively trivial. It may, however, provide very important information if the impacts of scenarios or policies are investigated which are defined by ten or twenty premises which also may be very different for the policies to be compared.

```
   EVAL 4;
    TIME OF INTEREST ?
   NFUT

    VALUE      !WT. !CF-!CF+ !  AGGREGATED EVALUATION (AGGEV)
    _________!___!___!____!____________________________________________________
             !   !   !    ! -250 -200       -100        0         100        200  250
    _________!___!___!____!!!____!_________!__________!__________!__________!____!
1   HEALTH   ! 95!  0!  50!                          0+++
    WEALTH   ! 55! 94!  50!                         -0+
2   SELFREAL ! 80!  0!  75!                          0+++++++
    INTPREST ! 40! 94!  75!                         -0++++
3   LIBSUBSY ! 45!  0!  94!                          0+++++
    LIBSOC   ! 60!  0!  50!                          0+
    ECOLEFF  ! 65!  0!  75!                          0+++++
4   ECONEFF  ! 80! 94!   0!                      ----0
    EFFCTRL  ! 45! 94!  75!                          0+++
5   SAFEDAN  ! 60!   !    !
    SAFESUPL ! 70!  0!  75!                          0++++++++++++
6   FLXIB    ! 75! 94!  75!                          0++++++++++++++++++++
    CPINNOV  ! 75!  0!  75!                          0++++++++++++
7   SOCRESP  ! 60!  0!  75!                          0++
    RESPFUT  ! 80!  0!  75!                          0+++++

   ****** NFUT
```

```
   EVAL 4;
    TIME OF INTEREST ?
   MFUT

    VALUE      !WT. !CF-!CF+ !  AGGREGATED EVALUATION (AGGEV)
    _________!___!___!____!____________________________________________________
             !   !   !    ! -250 -200       -100        0         100        200  250
    _________!___!___!____!!!____!_________!__________!__________!__________!____!
1   HEALTH   ! 95! 50!  50!                        --0++++
    WEALTH   ! 55! 50!  50!                         -0+++
2   SELFREAL ! 80! 50!  50!                        --0+++
    INTPREST ! 40! 50!  75!                         -0
3   LIBSUBSY ! 45!   !    !
    LIBSOC   ! 60! 50!  50!                         -0++
    ECOLEFF  ! 65!   !    !
4   ECONEFF  ! 80! 50!   0!                       ---0
    EFFCTRL  ! 45!  0!  94!                          0++
5   SAFEDAN  ! 60!   !    !
    SAFESUPL ! 70! 50!  75!                         -0+++
6   FLXIB    ! 75!  0!  75!                          0++++
    CPINNOV  ! 75!  0!  75!                          0+
7   SOCRESP  ! 60! 75!  75!                          0
    RESPFUT  ! 80! 75!  75!                         -0++

   ****** MFUT
```

Figure 3. Scenario 2: decentralization of the economy and increasing energy prices. (a) Evaluation of consequences in the near future. (b) Evaluation of consequences in the medium-term future.

Furthermore the example presented illustrates how the impacts and consequences of policies are distributed over time (compare Figure 3). Positive and negative effects do not appear simultaneously. Furthermore, DEDUC permits a systematic investigation of policy-impacts during the different time periods and of their evaluation with differing time horizons,[9] and differing weights for the time periods.

The overall evaluation may change considerably if, for example, positive consequences are concentrated in the present and near future whereas the negative consequences are being concentrated in the medium-term and distant future, a tendency which is to some degree illustrated by Figure 3.

These and other techniques offered by DEDUC to evaluate and analyse the impacts and consequences of real situations and policies provide a tool for optimizing policy measures with regard to the basic needs of a society and the particular goals of the policy-maker. If the policy-maker defines his value system, both in terms of a value hierarchy as it was used for the study reported here, and in terms of his temporal preferences, DEDUC models can be used to experiment with different mixes of policy measures under different scenario conditions until a mix is found which produces optimal results.

NOTES

1. The beginning of a change towards an actor-oriented paradigm in social systems theory is documented in Geyer, R.F. and Van der Zouwen, J.: *Sociocybernetics*, 2 vols, Martinuus Nijhoff, London, 1978; and in the *Proceedings of the 5th International Congress of Systems and Cybernetics, Mexico City 1981.*

2. For a discussion of such problems in the field of social indicators for "quality of life" see: Gehrmann, Friedhelm: "Valid" empirical measurement of quality of life, in: *Social Indicators Research, 5* (1978), pp. 73-109.

3. For the orientation theory see: (a) Bossel, H.: *Bürgerinitiativen entwerfen die Zukunft*, Fischer Taschenbuch Verlag, Frankfurt, 1978. (b) Bossel, H.: Orientors of nonroutine behavior, in: Bossel, H. (ed.): *Concepts and tools of computer-assisted policy analysis*, Birkhäuser, Basel and Stuttgart, 1977. (c) Bossel, H., Klaczko, S. and Müller, N. (eds.): *Systems Theory in the Social Sciences*, Birkhäuser, Basel and Stuttgart, 1976. (d) Bossel, H. and Müller-Reissmann, K.F.: Simulation of the cognitive processes of policy Analysis, in: *Policy Analysis and Information Systems*, Vol. 3, no. 1, p. 1-25. Chicago and Taipei, 1979. (e) Hornung, B.: Qualitative systems analysis as a tool for development studies. With case studies of Peru and Mexico, in: Geyer, R.F. and Van der Zouwen, J. (eds.): *Dependence and inequality*, Pergamon Press, Oxford, 1982. (f) Müller-Reissmann, K.F. and Bossel, H.: Zur Simulation kognitiver Prozesse bei Entscheidungen, in: Klages, H. and Kmieciak, P. (Hrsg.): *Wertwandel und gesellschaftlicher Wandel*, S. 539-55, Campus Verlag, Frankfurt and New York, 1979.

4. An encompassing description of the application of orientation theory for the construction of value systems, their computer implementation, and use for simulations is given in: Bossel, H. (ed.): *Kognitive Systemanalyse, Abschlussbericht zum Forschungsprojekt "Verhaltensbestimmende kognitive Elemente im nichtnumerischen Modell"*, sponsored by the Volkswagen Foundation, ISP, Hannover, 1982, Parts I and II.

5. For a complete description of the program and the methodology see the literature given in notes 3 and 4.

6. The colon (: ECONOMY) indicates that the statement is true for the aggregate, the economy in general, but not necessarily for all of its

subsectors (subclasses). In that case it would be "ECONOMY".

7. The complete information about method, empirical data, experiments and results of this project is presented in BOSSEL, H. (ed.): Kognitive Systemanalysis, *op.cit.*

8. The precise meaning of the orientors is explained in the appendix.

9. Trommsdorf, G. and Lamm, H.: An analysis of future orientation and some of its social determinants, in: Fraser, J.F. and Lawrence, N. (eds.): *The study of time: II*, Springer Verlag, New York and Berlin, 1975, pp. 343-61.

APPENDIX

Abbreviations used for orientors:
(in the sequence in which they appear in the diagrams)

HEALTH state of health of the population.

WEALTH general wealth and welfare of a society.

SELFREAL "self-realization" of the individuals.

INTPREST "international prestige" of a nation.

LIBSUBSY "liberty of subsystems": possibilities of the societal subsystems running their own affairs free from external constraints.

LIBSOC "liberty of the society": possibilities of action of a society both with regard to its own internal affairs and at the international level.

ECOLEFF "ecological efficiency": efficiency of technical, economic and social processes measured in terms of their use of physical resources, not in terms of money.

ECONEFF "economic efficiency": efficiency of economic processes, etc., measured in terms of money.

EFFCTRL "efficiency of control": control and steering capabilities of the state.

SAFEDAN "safety from dangers": physical security of the population.

SAFESUPL "safe supplies": safe sources and supply lines for important goods.

FLXB "flexibility": possibility of (structural) change of a society and its subsystems.

CPINNOV "capacity for innovation": creativity, inventiveness.

SOCRESP "social responsibility": responsibility for other human beings and societies in other parts of the world.

RESPFUT "responsibility for the future": responsibility for future generations and the living conditions they will find.

A critical review of the systems analysis approach to scientific prospectives and human needs

A. Danzin

The meeting organized by the International Institute for Applied Systems Analysis (IIASA) in September 1981 on the setting up and application of global models gave rise to a very interesting paper by Professor E. Pestel.[1] Professor Pestel was, with Professor Mesarovic, the builder of a world model which enabled the Club of Rome to write a report,[2] and allowed several forecasting scenarios to be worked out for use by various governments. In a rather touching way, Professor Pestel explains that after this accomplishment he was made Minister of Science and the Arts of the "Land" of Basse Saxe and that, in his new post, systems analysis was of no use at all to him. The ministerial function is trapped in a system of constraints favouring the short term, and influenced mainly by political irrationality. In such conditions the decision-maker has no time for an instrument which is too logical to integrate subjective factors and too prone to give a description—an unconvincing one at that—of the long term: for him, consideration of the less-than-immediate future plays a negligible role compared with the need to settle more immediate problems.

Critical of the application of systems analysis in the area of political decision-making, Professor Pestel is, in reality, criticizing the method of government itself in the circumstances in which he found himself.

It is, of course, by no means evident that management by the handling of current affairs only provides the best solution from the point of view of those administered to. Surely the fact that the long term is systematically sacrificed to the short term constitutes a fundamental criticism of decision-making methods rather than a condemnation of the systems approach and the use of scenarios of the future?

In any case, Professor Pestel cannot be taxed with being prejudiced against modelling, and for this reason his testimony is all the more important. For the same argument is often heard in the West, and more particularly in the mouths of industrial leaders. The industrialist too gives priority to his end-of-exercise balance sheet, and to that of the exercise immediately following. His horizon is generally limited to 2-3 years, or 5 at the most. He has to respond and adapt as best he can to sudden or seasonal fluctuations in the market. This tension, which lays stress on immediate performance, prevents him from taking into account, and trying to manage, the major trends which constitute the future. In a free market economy, only a few powerful companies—generally multinationals—are able to build a strategy aimed at replying to forecasts established for the next decade, and influencing them. These companies have their own study tools and, naturally, wish to protect their business secrets. They are therefore reluctant to share their accumulated knowledge on the exploration of the future with outside partners.

One can thus assert that the enthusiasm for systems analysis and its applications which marked the years 1965-70 has changed to disinterest, or even mistrust. It is impossible, obviously, to make such a statement and to leave matters at that, since the handling of current affairs on a day-to-day basis can only lead to catastrophe.

The population of the world will expand from 4000 million to 8000 million between 1975 and 2025, having already quadrupled during the last century. It is therefore necessary, in the space of these 50 years, to build as many factories, homes, schools and hospitals, to open as many mines and develop as many new agricultural projects, as exist today—and all this without managing to overcome the almost unbearable inequalities of wealth which are currently present. This will all have to be done while adjusting to an entirely new technological environment. Who can seriously believe that such an effort can be carried out in a climate of permanent improvisation? Decisions taken in 1982 will have noticeable results beyond the year 2000. How can one make the right decisions without the enlightenment derived from forecasting instruments?

SYSTEMS ANALYSIS AT GRIPS WITH COMPLEXITY AND UNCERTAINTY

Unfortunately, the situation with which we are faced cannot be analysed simply by putting the blame on decision-makers for neglecting the long-term view. Systems analysis, in itself, is contested as a tool for reading the future, and one must acquire a thorough understanding of its inadequacies and limits, as well as the fields in which it is useful.

The principal obstacle arises from the fantastic increase in the complexity and interdependency[3] affecting the most advanced societies. This increased complexity can be measured: it is shown by the information exchanged and turned over by the bureaucratic machine. We know that in the United States of America the professions which specialize in dealing with information, and the written and spoken word, i.e. banks, insurance companies and administrative services, today employ more than 50% of the workforce as against 15% at the turn of the century. In France, in the Paris region where the headquarters and major services of both public and private organizations are centred, more than two-thirds of those employed deal only with figures, letters or symbols relating to products or services with which they will never have any actual physical contact. In the field of science and technology the volume of publications is doubled every year. This means that during one period of 12 months some time between 1985 and 1990, the amount of scientific and technical/technological papers produced will be comparable—in quantity if not in quality—to the total contents of our libraries amassed by scholars and engineers from the beginning of the world until around 1940. As for the flow of information between the various countries in the temperate zone of the Northern Hemisphere, its volume has been increasing by 25% annually since the 1970s without any sign of slowing up beyond 1980. This flow covers all kinds of human activity: commercial and financial transactions; organization of overland, sea and air freight; aeroplane and hotel reservations; the battle against delinquency and violence; the exchange of artistic and cultural programmes; statistical data concerning social and economic life; scientific and technical information; remote control of industrial operations; and so on.[4]

In a world so interconnected, a decision taken with one particular goal in view can have repercussions on a whole range of other things. It produces not only direct, foreseeable effects, but also indirect ones that are often unexpected or contrary to good sense. The examples of such side-effects observed over the last few years are numerous,[5] and amount to almost unavoidable obstacles. The medical field is a particular case in point: it produces irrefutable benefits for mankind yet, these can also turn into calamities (I can cite as examples the constantly rising costs of health care that are difficult to meet; the overpopulation in countries of the Southern Hemisphere (the result of increased care for mother and child), the non-renewal of generations in the Northern countries (thanks to the efficiency of modern contraceptives), dismay at desperate therapeutic methods now used, the probability of increase of genetic illnesses as a result of the prolongation of life beyond the usual fertile period of those bearing chromosomic abnormalities). In numerous cases the very complexity of these phenomena prevents their being understood and, in consequence, being controlled. This is also the case in the monetary, economic and financial fields where methods of intervention seem lost in the face of our total lack of theoretical tools to understand the simultaneous appearance of inflation and unemployment and the violent fluctuations in exchange rates.

The fact that the biosphere can be perceived today as a system with finite limits adds complexity to the feeling of suffocation. Decisions bounce, so to speak, off the walls which demarcate the finiteness of the world. The Mediterranean must no longer be considered as an infinite ecological system, with an unlimited capacity for recuperation but, on the contrary, as a fragile system, intimately linked to the activities of those living around its shores.

The validity of the notion of the system, the global principle and the inseparability of the part from the whole is thus confirmed, drawing us away from Cartesianism and justifying a systems approach. At the same time, however, we are seized with a kind of vertigo when faced with the problem of modelling such a powerfully complex reality. By means of cybernetic reasoning we become aware of the fact that human society is, by its very nature, ungovernable and that, in order to control all its tangled parameters we would certainly need a tool of superior complexity, whereas we can only act from the inside, i.e. relying on a lesser tool.

In other respects, we know that human evolution obeys the laws of wasteful systems.[6] A given organizational structure interacts with the environment in which it develops. This interaction brings about changes in the environment, to such an extent that the structure is no longer adapted to the new conditions around it. We thus come to a point where the structure itself has to change. We know that there are only a small number of new, equally likely solutions. The tip of the balance towards one solution or the other is the result of a fluctuation occurring at the right moment. This fluctuation is, by its very nature, random. There is, therefore, in social as in biological life, an organic discontinuity in respect of time. There is also a basic principle of uncertainty which goes against all logic of forecasts. How, in these conditions, can we model a credible projection of the future? The pure and simple extrapolation of the past beyond the present does not conform with the laws of evolution; the principle of uncertainty stands in the way of our building convincing scenarios.

Such are the theoretical arguments against the systems approach that have been gathered together over the last few years. *In the eyes of many, they constitute its death sentence.* Other objections are raised, less serious from the theoretical point of view, but practically important. They are concerned with, in particular, the mediocrity of the statistical tool. The initial data are mostly suspect or non-existent. How can we foresee the future if we are not even capable of measuring the past?

THE REFUSAL OF OBJECTIVITY, THE "fixisme-transformisme" DIALECTIC

These criticisms are mainly well founded but, in reality, hide other, less worthy considerations. We are at a critical point in the history of mankind, but this prospect disturbs most of our contemporaries and reactions vary according to different schools of thought, and the cultural factors which affect them.

Specialists in the Exact Sciences, and engineers, base their entire system of thought on factual observation. Facts, whether resulting from real, measurable experience, or from confirmed theoretical explanation, constitute inviolable constraints for this category of thinker; they are within the field of objectivity. Action must respect these constraints. Thus engineers, when they build bridges, make use of the laws of gravity. Those minds trained in science and technology will never be tempted to deny the laws which are derived from measurement, number, or from the statistical observation of physical, economic or social phenomena. Anxious to define the valid areas of relations governing the behaviour of a system of parameters, they accept as true the interpolations within these valid fields, and as probably—with a calculable approximation—the extrapolations beyond the proved limits in this field. For them, the building of a model is a simplified but illuminating representation of reality, and the results which they draw from it are a source of reflection to enable them to ask the right questions of the system being studied. As for time, that towards the past accounts for the experimental evolution of the system as observed, and that towards the future makes it possible to describe the most likely events in unchanged structural conditions. If the forecast is unacceptable, catastrophic or unlikely, men trained in this school deduce that structural mutations are necessary and that they must be studied right away.

In our present-day society, these kind of thinkers act as forces of change, or as mutants, within the continuing evolutionary forces which have marked the transformation of living beings throughout their long history.

This kind of behaviour is an exception; and specialists must be aware of this. The recent meeting in Paris,[7] organized jointly by Unesco and the European Academy of Science, Arts and Literature, showed quite clearly that another part of the intelligentsia did not share this concern to give priority to factual constraints and to objectivity. Creative people, both literary and artistic, and philosophers and political thinkers, whose influence on contemporary society is equally important, do not believe in numbers, measurement, or mathematical modelling. They are the defenders of subjectivity; they embody the cultural forces which tradition has slowly formed; they favour intuition and sensitivity as opposed to calculation and logical reasoning. If the facts are contrary to the idea of reality which they receive from their mental model, then it is the facts which are wrong, especially if they are concerned with the future, which everyone knows is unpredictable. In our society, which is changing all the time under the effects of technological progress, these kinds of thinkers—in spite of what they imagine—are in fact forces opposed to progress, attached as they are to the protection of cultural identity and refusing to admit, as Schumpeter put it, that "development is a creative destruction". Upholders of creation, destruction makes them feel uncomfortable *(leur confrot ne s'accommode pas de la destruction)*.

Of course, if the literary and artistic intelligentsia—those who believe they are "culture" or "spirituality" incarnate—are so far removed from scientific thought, what can be said of those whose political and economic responsibilities lead them to concentrate on action and on the conduct of life today which has to be carried out before worrying about life tomorrow? What can be said about those who have the power in the media—press, radio and television—generally with little training in subjects relating to science and technology. And what can be said of the man in the street, upset in his working habits, worried about a future which he does not understand and disturbed by the flood of contradictory information which assails him.

The change-versus-sameness struggle belongs to the very nature of things; sameness is good, otherwise society would be entirely unstable; change is necessary, otherwise there would be no adaptation to the changing conditions of the human universe.

The defenders of systems analysis probably did not pay sufficient attention to this indispensable dialectic between the forces of faithfulness to culture and the forces of change. They did not realize that their enthusiasm was leading them to isolation, and that they were forming a separate group cut off from a vital part of the communications which they wished to maintain with their partners in order to help them to master the future. Nothing therefore would be more useful—and this could be one of Unesco's tasks—than to bring together men trained in different fields to think, not with a view to reaching a shaky compromise, but so that the necessary dialectic between the two basic attitudes may find a meeting ground for their expression and development.

What has just been said applies to the observation of the behaviour of partners in the most developed societies. The clash of culture and technology is, of course, even more brutal in the poorest countries, where scientific knowledge is imported or technology "transferred". This is a second subject for study. Today, we can no longer separate science from nature or from the humanities, culture, communication or education. This remark has consequences on the very structure of Unesco at a time when the medium-term plan is being studied. If, for convenience, it is necessary to keep to the tradition of vertical sectors, it would be desirable to envisage trans-disciplinary operations which would oblige men trained in different traditions of thought to carry out chosen projects, thus producing a confluence of ideas coming from different sources, which up until now—in order to protect the "purity" of each—refused to mix their waters.

One cannot insist too strongly on the need to establish the means for this joint effort. Those who have been instilled by the systems analysis of world problems no longer have the same overall view of mankind as other people. They know that in a very interconnected world no nation, however powerful it may be, can hope to protect itself against the spreading of an economic and social decline in the rest of the world; they measure, almost physically, the weight of solidarity.[3] They know that steps taken today to allow 8000-10,000 million people to live tomorrow are not enough. They attach the greatest importance to new factors such as "geotechnology" which is an offshoot of geopolitics, while recognizing the prime importance of the new centres of technological innovation, such as California and Japan, which have become "civilizations of communication". Their thought systems no longer rely on the Cartesian model but on the vision of an internal cybernetic of human societies of which all factors relating to possible action must be taken into account simultaneously in a global view. Believing—rightly or wrongly—that they have the truth, they have the impression that they are preaching in the desert.

IMMOBILITY AND ACTION

For those who have lived through the period running up to the Second World War, the present immobility is indeed reminiscent of that which preceded the conflict. Hitler had written *Mein Kampf* and his plan was being carried out in successive stages according to the published programme. Yet not one of his imminent victims had the willpower to act, while there was still time and at the least possible cost, by coordinating their action with that of the other powers suffering the same threat. The reaction came too late, and even though it was victorious in the end, immense sacrifice was caused.

The world situation today cannot be analysed with the same simplicity, and fortunately there is no powerful state whose explicit intent is to deprive its neighbours of their freedom. Nevertheless, peace is far from being consolidated, due to the immense heterogeneous factors which can be seen in the development of the different regions of the world.

The state of economic and cultural subordination of those zones which have come to experience the most acute demographic growth does not constitute stability—in fact, it does just the opposite. Scarcities are apparent in the field of nutrition and, there again, there is lack of capital for health, education and industrialization. The most fragile political zones are often those where the main sources of supply of rare metals and energy are concentrated. The international division of labour is developing in a random manner liable to send regions considered until now as prosperous havens hurtling towards under-development. The immense increase in productivity which will come with the spread of informatics and robotics will drastically reduce the number of jobs, and it is difficult to imagine how they will be compensated for by new activities. All these factors help to create a state of tension, the drastic solution of which could be armed conflict with uncontrollable repercussions. All these problems, seen as a whole, are part of the lot of mankind. They constitute part of the universal human heritage. Those who have studied them know that no geographic region is protected against their harmful evolution should this occur. Never has it been so necessary to study world problems as a whole.

Confronted with all the signs announcing imminent storms, however, what do we see? Opposition to progress, a refusal on the part of mankind to try and take its destiny in hand, and the incredible illusion that a good dose of national egoism will ensure an escape from the catastrophe.

Unfortunately, there are numerous examples showing this inertia. After the oil shock in 1973 the United States of America mobilized only a very small part of its immense means to diversify their energy sources and to combat the enormous waste of energy to which North American society has become accustomed. European industry seems incapable of reacting to the displacement (to the advantage, especially, of the Far East) of whole branches of industry, and allows Japan to lead the field in electronics and informatics, even though this is a major field for the global productive capacity of production, distribution and organization systems, and the most important technological element for the internal transformation of society. The "Communication Civilization" is indeed destined/called upon to replace the Industrial Civilization, thereby creating a change the like of which the history of mankind has never known, except perhaps for the passage from nomadism to sedentary agriculture.

And what can be said of the problems facing the developing world? How can one explain the lack of in-depth, global studies on nutrition, migration, urbanization? On agriculture alone, the experts differ fundamentally. For some, nearly all potentially arable land is already cultivated; others believe that nearly 50% remains to be conquered, over and above that cultivated at present. For some, the yield per acre can still be substantially increased; others regard excessive productivity as being the cause of irreversible soil destruction.

Education plans have failed. Illiteracy increases in volume; in those areas where it is decreasing the result is a massive migration towards the towns. Enormous cities of several tens of millions of inhabitants are being built which, as can be seen already, will be completely impossible to live in. Powerful migratory trends, both internal and external, are in progress; the ways in which they will develop are unknown, as is the way to control them in order to make the most of them.

The ecologists who declare that "Nature is threatened" are an active opposing force to the new techniques, and the movement is spreading over the various continents. Very few scientific studies are available to distinguish the physically and biologically sound arguments from those derived from a somewhat irrational and fearful imagination

Instead of mobilizing the best experts in an effort to understand the problems, making use of all the new methods of statistical analysis, data collection, storage and handling, assembling specialists in the physical sciences and the humanities to improve our understanding of these vitally important problems, and lighting the way for political decisions with an objective description of the major trends the consequences of which can be forecast (and all of the above-mentioned could be accomplished with a few thousand million dollars), mankind does not hesitate to spend one hundred times more each year on armaments which are already more than powerful enough to destroy the whole biosphere.

Even if the controversy over the validity of the application of systems analysis is only partially justified, the systems approach constitutes a positive approach to all these problems. It allows us to ask relevant questions in almost quantitative terms; translates thought into concrete terms, to be shared between scientists, men of literature, engineers and sociologists, technicians and specialists in matters of the spirit and ethics.

But an obstacle stands in the way of such a dialogue; and that is the lack of understanding caused by complexity.

AN ESSENTIAL STUDY: HOW TO MASTER THE COMPLEXITY

The study of power relations in a complex society has yet to be undertaken.

Those with responsibility are only now becoming aware that human society, in becoming more complex, has crossed a threshold that radically changes the nature of power. A change in the structure of consultation and hierarchical organization becomes necessary, within human groups, businesses, administrations, regions, as in the governing of states and in international relations.

The conduct of business affairs must preserve this complexity without violating it by simplification.[8] This statement is the result of observed experience: this is how those who have chosen to devote their economic efforts to informing bodies and individuals who carry out the orders, and to the search for a general consensus in favour of technical and social innovation, have best succeeded. But this manner of understanding action is also in line with the theoretical ideas which we can have on the mechanisms of biological evolution, of which the peak today is human society, i.e. a permanent surge towards the accumulation of information, and more and more communication, complexity and awareness.

Organizations as such must be reconsidered in order to adapt to this complexity.[9] One of the best syntheses on the subject was made by Jean Voge under the title "A new economic order in information and communication".[10] Jean Voge bases his work on the writings of three Nobel prize winners—Herbert Simon, Ilya Prigogine and Manfred Eigen. He shows how new paradigms could be built which would allow the inevitable trial-selection mechanisms to be respected while relying on the occurrence of phenomena of chance. In other words, these concepts lead us to practical organizational formulae which would correct to a large extent the basic uncertainty and unmanageability which we are obliged to recognize as objective data of the evolution of human societies.

We must not harbour any illusions about the difficulty of the complexity problem. A great deal of controversy and opposition—much of it emotional and irrational—will be heard before the matter is clarified and practical conclusions are drawn from the studies. Time is needed. That is why the means for deeper consideration should be provided as soon as possible. This could be the subject of a study sponsored by Unesco, but carried out by already existing centres such as (to limit oneself to Europe) IIASA in Vienna or the FAST group in Brussels. It would also be useful for a scientific focus—or network--to be set up for this type of research in the Southern hemisphere in order to help developing countries obtain for themselves an improved and more objective picture of the efforts which they themselves will have to make.

NOTES

1. Pestel, E.: Global modeling at the service of the policy maker. Conference held by IIASA, September 1981.
2. Mesarovic, and Pestel, E.: *Mankind at the turning point (Strategie pour demain,* Le Seuil, 1974).
3. See the work of Jacques Lesourne, especially *Les Mille Sentiers de l'Avenir,* 1981.
4. See the Anderla report on Scientific and Technical Information, OECD, 1974, and other OECD reports on the flow of data across frontiers, 1977.
5. Boudon, R.:

6. See the work of Ilya Prigogine, especially *La Metamorphose de la science ou la Nouvelle Alliance*, 1980.

7. See a report of a workshop in Paris, organised in November 1981 with the help of Unesco (in preparation).

8. See the work of Michel Crozier.

9. Serieyx, Hervé: *Mobiliser l'intelligence de l'entreprise*, E.M.E. edn, Paris, 1982.

10. Voge, Jean: *Telemates, mediateurs et microsocietes*, unpublished study: see IDATE, Montpellier, 1982.

Methodological problems of science and the evolution of human needs

J. M. Gvishiani

INTRODUCTION

The theme of our symposium—"Scientific Forecasting and Human Needs"—is an attractive one due to its urgency and comprehensiveness. It is urgent owing to the fact that it concerns actual human needs—those of individuals, groups of people, nations, and mankind as a whole. As far as we scientists are concerned, it is doubly urgent since it covers the issues of scientific development perspectives, the essence and lines of scientific community activities.

On the other hand, the theme is as comprehensive as life itself, for it does not embody separate scientific disciplines but science as a whole. What is more, it does not concern science *per se* but its relationships with human needs which are also treated as a complex system in all their variety and with due regard to their interrelationships—with one another, science and society as a whole.

The urgency and comprehensiveness of the theme makes the analysis of its methodological aspects very desirable indeed. I have had several opportunities of expressing my views on some of them, including Unesco conferences and publications.[1] Accordingly, in this paper I will touch upon the following issues: (1) specific interrelationships between science and human needs in an environment marked by scientific and technological progress; (2) methodological problems of systems approach to the needs; (3) diversity of national experiences and necessity of international scientific cooperation with a view to meeting human needs.

SCIENCE AND HUMAN NEEDS IN AN ENVIRONMENT OF SCIENTIFIC AND TECHNOLOGICAL PROGRESS

The present epoch is highly dynamic and multifaceted. It is often referred to as the age of space and atomic energy, ecological crisis and holocaust, population explosion and downfall of colonialism, antagonism of the two social systems—capitalism and socialism—and the epoch of struggle for international cooperation, safeguarding peace and harnessing the arms race.

Our epoch is also referred to as the time of scientific and technological revolution, and with reason. Whatever facet of the present age one turns to, he will see that science plays a notable role in its formation.

Scientific and technological advance represents a qualitatively new level of productive force development distinguished by the transformation of science into a leading productive force. It changes all the material and technical bases of production in line with the revolutionary progress of science and technology. One of the key lines of advance is industrial automation, widespread application of electronics, computers and microprocessors in controlling technological systems. This radically changes the position of man in production, and gradually in non-production, processes.

In the final count the scientific and technological progress is not only the unity of drastic changes in science and technology but also the transformer of the pattern of needs and values of individuals and social groups, nations and civilizations. Though in different societies there are different social premises for scientific and technological progress, and often opposite social implications, it results in profound transformations in any one of them.

Hence, the relationship of science and human needs in conditions of scientific and technological advance should be considered in a wide social context. One of the approaches is presented schematically in Figure 1.

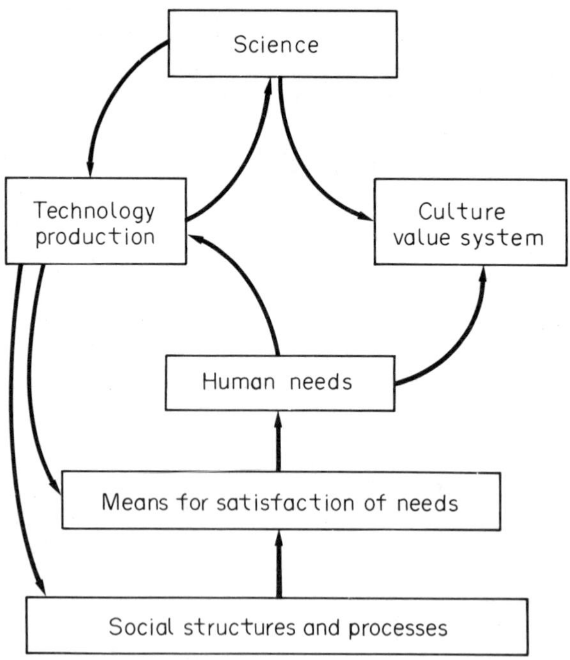

Figure 1. Science and human needs in the social context (in the environment of scientific and technological progress)

As can be seen, there are two groups of factors behind the relationships of science and human needs: first, technological including products or means for satisfying the needs; second, social and recreational, involving social structures and processes, culture and value systems. The presence of these factors makes the relationship between science and human needs truly dialectical: the achievements of science translated into the advanced processes act as a means of meeting the human needs and at the same time as a source of new demand.

On the basis of past experience we know that the above groups of factors have different impacts upon science in different social and cultural settings and at various stages of development of science itself. The technological factors always bear on science. Their influence is rather strong and is realized through orders and investments. At the same time there is an ever-growing reciprocal influence of science on production. As was previously mentioned, science turns into a leading productive force. As for the social and value factors, the intensity and nature of their influence depend on the type of society and its culture, as well as on the stage of development of science as a social institution.

In this connection, one may ascertain substantial changes in approaching the social function of science. Until fairly recently there prevailed the ideology of "pure" science, which develops spontaneously from discovery to discovery, independent of social factors or needs. Then emerged the concept of the technological and economic function of science as a direct productive force intended for the fulfilment of social orders external to it. Nowadays, there is growing realization of the need to direct the development of science and the generated technological advance to clear humanistic goals. The complexity and controversy of the processes encountered by mankind in the course of scientific and technological revolution, the high rate of the processes and scale of implications compel society to treat science as a social institution assigned with responsibility for preparing, validating and forecasting strategic decisions.

It must be noted that such perception of science was first established in socialist countries. It is quite understandable as the public ownership of the means of production makes not only imperative but feasible the assignment of extensive, socially significant problems and objectives to science. In the Soviet Union we make every possible effort to direct the development of science towards the satisfaction of human needs, and the full utilization of science potentialities for the most rapid implementation of the nationwide programmes, such as the food programme. The characteristic of the current stage is that the posing of problems and definition of objectives related to human needs, increasingly becomes the job of science proper. In his report to the 26th Congress of the CPSU, L.I. Brezhnev stressed that science itself should be a constant "troublemaker", pointing to the areas where there are signs of stagnation and backwardness, where the present level of knowledge could secure faster and more successful advancement. Thought should be given to ways of turning this work into an integral part of the management mechanism.[2]

There is a similar trend in the West. This is realized, for instance, in numerous proposals and demands concerning the humanization of science. The majority of them originate with more realistically-minded scientists, who have realized the futility of polemics between the advocates of scientism and their opponents—technocratic optimists and despairing pessimists—and are seriously engaged in the search for constructive ways of problem solution. The very demand for humanization of science implies that in the long run science should strive towards humanistic goals. Accordingly there are suggestions for wider dissemination of information on the possible impact of scientific advance, for more extensive communication between scientists and policy-makers responsible for decisions with respect to scientific development, for a stronger influence of the scientific community on shaping the researchers' ethics, and so on. In short, it is the question of coordinating the efforts of researchers, scientific communities and politicians in mapping out the ways of scientific, technological and social progress.

The viability and bright prospects for humanization and socialization of science require that they correspond not only to the needs of modern society's progress but also to the immanent needs of the development of science proper; the social content of scientific labour as labour in general. The product of this labour—knowledge—is also of universal, general significance. It is knowledge obtained by men for men. The aggregate human need is at one and the same time a point of departure and the destination of research. The humanistic orientation of science is not its external but internal orientation, i.e. self-orientation. Its development testifies to a high maturity of both science and culture, and entire society.

The realized, and not simply declared, orientation of science towards human needs is a complex, multi-level hierarchical process. It involves: firstly, multiple option forecasts and estimates of global, regional and national development which allow one to set goals for science as a whole; secondly, analysis of likely directions in the evolution of human needs in the medium- and long-term perspective, formulation of new evaluation criteria more adequately reflecting the changing relationships; thirdly, study of activity sectors and groups of needs which make it possible to shape the goals and objectives for scientific areas and disciplines; and fourthly, the drawing up of comprehensive goal-oriented programmes and the introduction of innovations providing for the practical application of the available scientific potential to meet the specific human needs.

Great significance is attached to the executive influence on scientific development on the national level. The specific content of the influence, or science policy, involves setting the scale of research and development (human and material resources), R&D decision rules, and assignment of R&D priorities. All the influences pursue two major groups of goals: scientific support of the nation's development and the balanced evolution of science itself in the entire diversity of its cognitive, social and cultural functions. At the same time the valid formulation of the goals presupposes more general targets taking account of the processes and tendencies of regional and global nature. On the other hand, the national and sectoral goals and objectives are accomplished through a totality of programmes and innovations providing for practical creation of means for the satisfaction of human needs and evolution thereof. It must be kept in mind that on this level the executive influence is not directed at science as a whole and, as a rule, not at individual scientific disciplines but at interdisciplinary relationship of a certain number of sciences. It makes the problem of science orientation at human needs satisfaction much more complex.

METHODOLOGICAL PROBLEMS OF SYSTEMS APPROACH TO THE NEEDS

The consideration of science and human needs relationships, and science orientation towards human needs, presupposes a more extensive study of methodological problems involved in research on the needs proper.

First of all, there is the question of the type of methodology most adequate to the nature of the system under investigation. It seems reasonable to single out the following two characteristics of human needs as the object of research. In the first place, it is a complex, multi-faceted system within the competence of numerous disciplines belonging to substantially different fields of knowledge. Secondly, it is rather a practical object whose study should not be confined to purely academic exploration. It is necessary to ensure the most extensive application of the obtained theoretical results to the practical activities of society, its institutions and organizations.

We believe that the studied object, with its specific characteristics can well be subject to systems research as the latter has proved its effectiveness as applied to a host of complex problems. In the first place, this methodology is based on a systems approach which is part of the general science methodology. This implies its orientation towards science as a whole and not towards any individual science or group of sciences, towards an integration (naturally, from a specific point of view) of advances in social, natural and physical sciences as well as of the practical experience, primarily in the area of organization and management. Then the general scientific nature of systems approach stipulates its particularly close relationship with philosophy. Without claiming that its conclusions have philosophical value, the systems approach acts as one of the links between the general philosophical methodology and the one of specific sciences.

Yet another characteristic of the systems methodology is its direct relevance to complex real-life problems. Special significance in this connection is acquired by applied methods which materialize the integrative capabilities of the systems approach and ensure interdisciplinary synthesis of knowledge of the studied processes aimed at the solution of theoretical and practical problems.

A systems approach to the problem of human needs requires their interpretation as a component or subsystem of human vital activities. The specific function of this element is that it makes the entire system of activities dynamic and determines the direction and force of the activities. It must be kept in mind that different human needs are closely linked and form an integral system governed by specific laws and mechanisms of its origination, functioning and evolution. One of the laws is the transformation of needs from elementary to sophisticated ones, from absolute domination of material needs to a steadily growing weight of spiritual needs.

Another law relies upon the growing dependence of the individual need satisfaction on the extent and pattern of satisfaction of collective needs (of groups, societies, humanity). This law becomes particularly evident in an environment of scientific and technological progress marked by the emergence of new collective needs extremely significant at all levels of the human needs system. For example, one of the most urgent is the need in harnessing the arms race, the prevention of wars, and safeguarding peace on the globe. All the other needs will not be realized, in the final count, unless the aggregate problem of meeting the foregoing crucial need is solved.

We are witnessing the growing significance of the need for an environment—both natural and urbanized—that man deserves. One should not infer from this, however, that the needs on the individual level, or the immediate needs of each human being, become less significant. On the contrary, all of us know that in many regions of the world the problems related to the satisfaction of human needs for adequate food, clothing, housing, health care and education become more and more acute. Even in industrially developed countries millions of able-bodied citizens cannot satisfy the human need for creative work.

Hence, the question is not of the reducing significance of such needs but of the growing dependence of their satisfaction on the extent and pattern of satisfying the collective needs, i.e. the ones every individual is connected to indirectly through some or other social collective—group, society or mankind; in other words, the question of the growing complexity of human needs as a system. The study of this system requires a wide range of systems analytical tools. One of the effective tools for investigating such complex and diverse problems as human needs and ways of their fullest satisfaction is computer simulation. This technique is being developed in a number of Soviet research institutions. A substantial contribution to its development is being made by the specialists of the All-Union Research Institute for Systems Studies (VNIISI) of the USSR Academy of Sciences by turning it into systems modelling. I would like to emphasize that the term "systems" is not used here as a fashionable epithet but as a means of accurately describing what has been brought into the development of simulation methodology.

The essence of the methodology is the interrelationship between quantitative and qualitative research techniques, advantages of formal computer modelling and informal capacities of man, which adequately accounts for the complexity of the object and problems under study. This is accomplished primarily by way of constructing such a formal simulation system that originally does not contain a model but is just a tool for enabling the researcher to generate quickly the desired model structure or system of models conforming to the goals of research, to specify the system with quantitative correlations describing the relationships between its elements, and to conduct research and model applications for a wide range of problems. The above-mentioned speed is made possible by automatic routine procedures throughout the modelling process.

By-passing a detailed description of the formal simulation system, I would like to point out that the construction and application of such a system not only belittles but substantially increases the significance of non-formal components on which the quality of designed models and obtained results ultimately depends in the final count. The detailed analysis of both the non-formalized system and non-formalized components of systems modelling was made by VNIISI experts and described in a number of publications.[3,4] I believe the VNIISI experts present at this Symposium will touch upon some of the results of the analysis in the course of the subsequent discussion.

The systems modelling methodology was elaborated in response to the growing problems of global and regional development. Since these problems are closely related to the problems of human needs, global modelling constitutes one of the most important areas of research into the problems connected with human needs.

I have already written[5] that the models by J. Forrester[6] and D.H. Meadows,[7] the Latin American model by Barriloche,[8] and other models of global development, studied the opportunities of meeting certain human needs. However, the set of variables accounted for in these models was far too small; also limited were the methodological principles of their design.

Applying the above methodology of systems modelling, the VNIISI specialists can consider a far wider range of variables relating to the problems of human needs satisfaction. An incomplete list, shown in Table 1, includes about 130 variables.

TABLE 1 List of Variables Related to Human Needs Satisfaction

Submodel	Variable
Macroeconomics (M)	1. Gross Regional Product.
	2. *Per capita* income index.
	3. Index growth rates.
	4. Accession rates in production sphere.
	5. Share of Gross National Product in production accumulation.
	6. Share of Gross National Product in industrial payroll.
	7. Share of Gross National Product in science.
	8. Share of Gross National Product in social and recreational investments.
	9. Index of education *per capita* expenses.
	10. Index of health care *per capita* expenses.
Demography (D)	21. Total population.
	22. Population density.
	23. Population growth rate.
	24. Age pattern.
	25. Share of young population.
	26. Young increment rate.
	27. Male (female) population.
	28. Share of rural population.
	29. Urban population.
	30. Urban population growth rates.
	31. Share of urban population.
	32. Able-bodied population.
	33. Share of able-bodied in population.
	34. Able-bodied growth rates.
	35. Mortality per 1000.
Food (F)	41. Level of nutrition.
	42. Famine.
Income differentiation (ID)	51. Gini coefficient.
	52. Gini coefficient variation rates.
	53. Relationship between the top 10% and the remaining 90% of population by income.
Labour (L)	61. Professional pattern, 6 professional macrogroups.
	62. Professional pattern, 'n' professional microgroups.
	63. Labour content in microgroups (level of monotony and mechanization).
	64. Working-class share.
	65. Share of low-skilled and unskilled workers.
Education (E)	71. Average educational level (average skill).
	72. Male average educational level.
	73. Female average educational level.
	74. Urban average educational level.
	75. Rural average educational level.
	76. Share of illiterate.
	77. Education development rate.

TABLE 1 List of Variables Related to Human Needs Satisfaction (continued)

Submodel	Variable
Strikes (St)	81. Number of strikes.
	82. Number of strikers.
	83. Number of strike man-hours.
	84. Number of unionists.
Spare time (ST)	91. Physiologically necessary time for males.
	92. Physiologically necessary time for females.
	93. Physiologically necessary time in town.
	94. Physiologically necessary time in the country.
	95. Time of domestic work with men.
	96. Time of domestic work with women.
	97. Time of domestic work in town.
	98. Time of domestic work in the country.
	99. Spare time of men.
	100. Spare time of women.
	101. Spare time in town.
	102. Spare time in the country.
	103. Recreational behaviour of men.
	104. Recreational behaviour of women.
	105. Recreational behaviour in town.
	106. Recreational behaviour in the country.
Scenarios (Sc)	121. Form of ownership of the means of production.
	122. Nationalization rates.
	123. Social security.
	124. Vertical mobility.
	125. Religion (type, uniformity).
	126. Racial and national heterogeneity.

The modelling system helps study the ways towards improved food and housing, accessibility of education and health care to the entire population, civil rights guarantees including the right to work, and a number of other universal problems such as conservation and restoration of the natural environment, securing peace and harnessing the arms race. The study of these problems is conducted both on a global and regional scale (currently, in nine regions of the world).

The preliminary analysis, conducted with expert judgment, helped identify the 20 most significant problems of world development (Table 2) and define their causal dependence on one another. The nature of causal dependence allowed a breakdown of the totality of problems into three large groups (Figure 2), notably: general system K-1, economic K-2 and social K-3.

Special emphasis should be given to the general system group K-1, which includes such problems as safeguarding peace and curbing the arms race, bridging the gaps between the levels of regional economic development, consolidation of people's effort towards solution of global problems, etc. There is close relationship between all the problems within the group, which in its turn exerts a strong determining influence on the two remaining groups: 22 types of influence on K-2 and 19 on K-3.

TABLE 2 List of World Development Problems

1. Safeguarding peace and curbing arms race.
2. Conservation and restoration of the natural environment.
3. Replacement of depleting production resources (energy, minerals, water, etc.).
4. Bridging the gap between the levels of regional economic development.
5. Elimination of dangerous and common diseases, conservation of gene pool of mankind.
6. Restructuring of international economic relations.
7. Greater role of international organizations (UN, Unesco, etc.).
8. Population food supply.
9. Accessibility to education and cultural values for the entire population.
10. Guarantees of human rights.
11. Accessibility to health care for the entire population.
12. Housing supply to the entire population.
13. Cutting down criminality, alcoholism, drug addiction rates.
14. Reduction of heavy and harmful types of work.
15. Securing balance between the population growth rates and *per capita* GNP growth rates.
16. High or low level of income differentiation between different social groups (within the country).
17. Social equality or inequality.
18. High or low level of public consumption funds.
19. Orientation at the most rapid or socially balanced economic growth.
20. Capitalism or socialism, or pluralistic society.
21. Consolidation or dissociation of mankind efforts directed at global problem solution.
22. Holistic man—scientifically based goal or spontaneous result of historic development.

The economic group, K-2, is the most weighty one in the hierarchical structure of problems. Subject to a certain influence of the general system group, K-1, it exerts its own influence on both of the groups, K-1 and K-3.

The social group is nearly completely determined by the other two groups. Its dependent position can be interpreted from two essentially different points of view. If a society develops to a large extent spontaneously (in market economies), then the solution of social problems comes about spontaneously in the wake of other problem solutions. In the case of the society's development proceeding on a plan-wise basis (in planned economies) the solution of social problems represents a conscious task while the governing factors should be treated as tools for optimizing these final criteria.

This dual interpretation also holds true with respect to the majority of problems relating to human needs, since these problems belong to the social group, K-3. Some of them, however, belong to the economic and general system groups. The universal need for peace dominates the entire structure of the world development problems (in Table 2 it is at the top of the list).

There are some methodological problems of systems approach to the needs. Having considered in brief the two aspects of the theme "Scientific forecasting and human needs" both from the point of view of science and the needs in their interrelationship, we shall proceed to the issue of variety of national experiences and the necessity of international cooperation in research aimed at satisfaction of human needs.

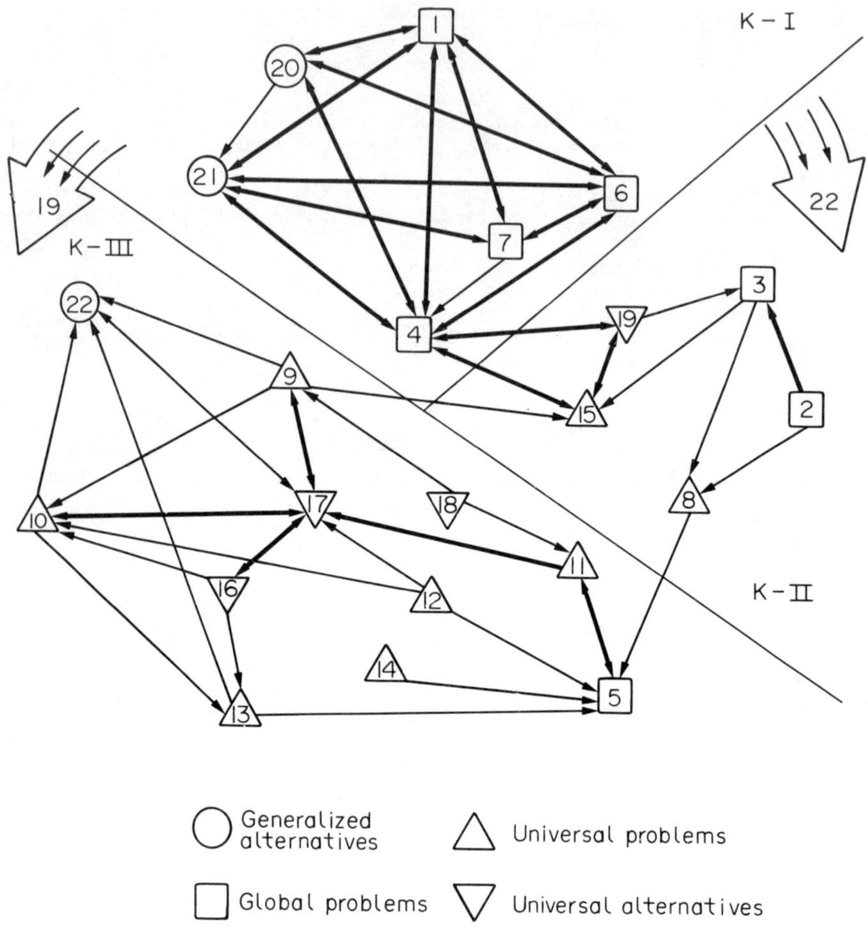

Figure 2. Three sets of world development problems and options.

DIVERSITY OF NATIONAL EXPERIENCES AND THE NECESSITY FOR INTERNATIONAL SCIENTIFIC COOPERATION

Considering the problem of relationships between science and human needs, between science and society, we cannot but confirm the growing conflict between the potentialities of science and its failure to materialize them for the benefit of humanity. Realization of this fact is enhanced by the stronger tendencies testifying to the emergence of a new social community on our globe—a qualitatively new-world civilization. Scientific and technological progress is one of the key driving forces in the development of this process. Due to advances in science and technology mankind is gradually becoming a global system wherein the processes taking place in the most distant geographical locations become interrelated. There is no doubt that the progress of science and technology leads to the growing ability of man to master the natural forces in his own interest. At the same time, however, it breeds considerable negative consequences of unreasonable application of scientific and technological achievements because of the imperfect social systems.

Hence, a stronger emphasis should be placed on the functioning of the global system, of the entire complex of international relations including those in the field of science and technology.

Serious failures in implementing the development programmes initiated by the international community, as well as the emergence and aggravation of global problems jeopardizing mankind, have led to the realization that the scientific and technological progress achieved in some countries cannot be mechanically transferred to others, and does not lead automatically to social progress.

It was very significant to see, during the 1970s, the limitations of the traditional view of the role of science in the solution of problems facing an individual and the whole of society. It was typical for the "modernization" concept. This approach was based in particular on the possibility of reducing the variety of development problems to a list of specific technological, economic and other objectives, and of applying to each one the achievements of science and technology.

The perception of the necessity to redirect science along humanistic lines, consistent with the economic and social needs of society, was accompanied by a reassessment of the emerging global problems as complex problems requiring a systems approach, the synthesis of natural, social and physical sciences.

This led to new forms of cooperation of scientists and specialists from all over the world. And here we must highly commend the activities of such organizations as the Club of Rome, the International Institute for Applied Systems Analysis (IIASA, Laxenburg, Austria), the International Federation of Institutes for Advanced Studies (IFIAS), which have concentrated on interdisciplinary problems of modern times. This trend towards a wider international cooperation for the purpose of accumulating the scientific achievements of many countries, towards integration of effort and concentration of resources for the solution of key problems, has become characteristic of the majority of traditional organizations in the last decade. No doubt this became possible only in the favourable political climate and detente of the 1970s.

Another important development was the Unesco initiative in generating a special programme with a view to promoting the definition of research priorities from the point of view of social goals and human needs. Soviet scientists actively contributed to the programme almost from the very beginning. We always believed, and still do, that the success of the programme can only be achieved through accounting for the diversity of national experiences, further extension and development of international cooperation. In our opinion, the value of this International Seminar held within the framework of the Unesco programme is, in particular, that it will consider different projects, elaborated with specific national approaches. The Soviet experience in the elaboration of the Comprehensive Programme of scientific and technological progress of the USSR is reflected in the presentations to be made here.

Speaking of the ever-growing significance of science in the development strategies, one should pay attention to the evolution of large-scale international and national projects that took place in the previous decade. If the first global projects initiated by the Club of Tome, and the large-scale national projects sponsored by various funds, could not integrate science and technology as an important factor of sócioeconomic changes, then the current projects are specifically oriented at evaluation of long-range perspectives of scientific development and its contribution to solution of the crucial current problems.

It now seems necessary to focus on the problem whose significance has grown considerably of late. It is the destiny of science in connection with the problems of war and peace, especially under current circumstances. The present-day tendencies cause some concern, and the emerging vicious circle is rather dangerous. Science is not only directly involved in the creation of new weapon systems, not only promotes the arms race, but is also the object of the corroding influence of militarization cultivating utilitarianism, narrow-minded approaches, dissociation of scientists, their isolation from the needs of the wide public, and their alienation from managing science and controlling the application of its achievements.

These baneful trends should be confronted by the active stand of scientists advocating science as the common property of humanity as a whole. First of all, there is the international cooperation of scientists. Scientific and technological contacts, and international communication of scientists are not only imperative for further advance in science but they provide a bridge connecting peoples in the face of common problems. Under the present circumstances, when scientific and technological factors underlie numerous problems, international scientific contacts are called to play a positive role in providing a basis for political negotiations.

All the above speaks in favour of a radical re-evaluation, on the part of scientists, of their role in society and in the world. They should renounce the positivistic approach to the concept of "objectivity", "impartiality" or, to put it another way, dissociation from the important processes taking place on the globe. We believe that enrichment of the traditional experience of science with the principles of humanistic orientation, and the striving for cooperation with representatives of other disciplines, are implemented through the study of scientific problems as applied to human needs. At the annual general meeting of the USSR Academy of Sciences, Academician A.P. Alexandrov, its President, stated the position of the Soviet scientists as follows: "Humanity faces large-scale problems; and we, scientists, properly perceive them and believe that it would be most reasonable to integrate efforts of all the nations of the world, all scientists in order to provide sufficiently favourable conditions for the existence of mankind."[9] The present Symposium illustrates the active stance of modern science in favour of extensive international cooperation.

NOTES

1. Gvishiani, J.M. Scientific-technical progress and the social goals of science. *Impact of Science on Society*, vol. 29, no. 3, Unesco, Paris, 1979.
2. *Proceedings of the 26th Congress of the CPSU*, Politizdat, Moscow, 1981.
3. Gelovani, V.A. Man-machine systems for modelling processes of global development (in Russian), in *Sistemnyje issledovani Metodologicheskije problemy*, Ezhegodnik, Nauka, Moscow, 1980.

4. Lapin, N.I. The function of non-formalized elements in modelling systems (in Russian), in *Nefomalizovannyje elementy sistemy modelirovanija*, Trudy seminara. VNIISI, Moscow, 1980.

5. Gvishiani, J.M. Global problems in the Year 2000 and the role of science in their solution, in *Science, Technology and Global Problems*, Pergamon Press, Oxford and New York, 1979.

6. Forrester, D. *World Dynamics*, Cambridge, Mass., 1971.

7. Meadows, D.H. *et al.* *The Limits to Growth*, Universe Books, New York, 1972.

8. Herrera, A.O. *et al.* *World Model*. IIASA, Laxenburg, Austria, 1974.

9. Alexandrov, A.P. Vstupitelnoje slovo na godichnom sobranii akademii nauk SSSR (18-20 March 1981). Vestnik AN SSSR, no. 7, Nauka, Moscow, 1981.

Methodological problems in forecasting fundamental research related to human needs

I. Malecki

FORECAST OF SCIENTIFIC RESEARCH

Over the past 30 years the forecasting of scientific research has attracted much attention in the scientific community and in science policy institutions. This results from the fact that science exerts an important influence on the shaping of reality and that its development is controlled by a specific order of logic. An image of the world of the future would be incomplete, and even distorted, without taking into account the development of science and its consequences, and without using specific methods to foresee this development. Science forecasting mainly concerns the future and its developing trends; that is, predicting the fields of research which advance most rapidly and the expected important discoveries. Assuming this kind of forecasting, one should answer the question: "What are the practical consequences of the expected discoveries?" Possible applications are sought amongst different and very diverse fields of practical activity. It is presumed that such a method is capable of forecasting most successfully possible developments in fundamental research. Trends in social and technological research should be forecast in another way: first, social changes and technical advances are forecast (e.g. by the Delphi method), and the findings thus acquired form the starting point from which to go backward and predict the results of earlier scientific research which could be the basis for related achievements in practice. This forecasting method is based on the assumption that social change and technological advances are controlled by their own order of logic while scientific research plays a subordinate role in the general development of society and is forcibly stimulated to an extent by practical requirements. Although such reasoning raises some objections, this method is rather extensively employed because it directly forecasts events of a high practical importance that are most interesting for the people to whom the forecasting is addressed.

Forecasting the development of science from the point of view of human needs is a new challenge, and requires the combination of both methods mentioned above. The problem involves two sequences of forecasting: the range and variation with time of human needs, and the sequence of scientific discoveries. The two sequences are treated as mutually dependent, which is a methodological assumption that affects the entire forecasting process. The feedback between human needs and the results of scientific research follows from the fact that scientific research creates the improvement of the material means indispensable for covering human needs and simultaneously science policy prefers—or should prefer to some extent—research oriented towards a more complete meeting of these needs. The growth of human needs follows from complex social processes which are only under control to a very small degree.

Apparently more flexible is the science policy which enables research priorities to be allocated in a rather arbitrary way. However, it should be noted that material means alone are not sufficient for scientific progress, because there is always the risk of undertaking research projects and other scientific creative activity the effects of which cannot be predicted and measured.

FORECASTING A SCENARIO OF HUMAN NEEDS

The forecasting of human needs is based on an assumed scenario, the main items of which are:

1. Demographic Forecasting

This determines the quantitative estimation of human needs, but in many cases quantity changes into quality; for example, traditional energy sources may prove to be sufficient for a low rate of population growth, while for higher rates it will be necessary to find new, revolutionary solutions for the problem of energy shortage.

2. Undisrupted Development

Forecasting generally assumes that the lives of nations and societies will not be disrupted by catastrophic disasters. Otherwise, as in the case of, for example, a world war or a dramatic change in the earth's climate, forecasting will have no sense because any foretold course of events based mainly on extrapolation is changed by unpredictable situations.

3. The Evolution of Needs

This evolution is both quantitative and qualitative in nature. If it is assumed that the number of a given population increases but that individuals become poorer and poorer, the needs not only drop relatively in quantity, but their very structure becomes more primitive. Needs related to starvation and elementary health conditions remain the primary problem. If it is assumed that the population develops in all directions, the needs related to civilization and intellect become of primary importance.

4. The Degree of Uniform Development

Forecasting must take into account the division of the worlds inhabitants into the rich and the poor, the producers and the consumers, the latter being dependent on the former in both cases. At present, the twenty most developed countries provide approximately 20% of the world's industrial output and possess about 80% of the world's potential in fundamental research. They are almost the sole initiators of technological innovations. The preservation of this imbalance, or a change towards its at least partial elimination, influences the distribution of human needs and ways of satisfying them in a radical way.

It is beyond the scope of the present paper to formulate a forecasting of human needs based on the choice of scenarios outlined above. Nevertheless, to illustrate the methodological problems encountered in forecasting fundamental research, we shall use a working hypothesis assuming that:

(a) the world's population will increase in accordance with the so-called medium demographic hypothesis;
(b) no catastrophic event will occur;
(c) elementary human needs will be met to a permanently increasing extent, and thus civilization and intellectual needs will be more and more important;
(d) disproportions in the distribution of production output and the potential of fundamental research will decrease but the major concentration of the centres of technological advance will remain relatively unchanged.

All the above factors indicate that both quantitative and qualitative human needs will increase on the global scale.

APPLIED RESEARCH FORECASTING

I would like to discuss now the way in which the growth of human needs will affect the forecasting of research.

One must, first of all, determine the role of applied research in meeting human needs and then point out the place of fundamental research in the whole process. Here, material and non-material needs should be considered separately. In the field of material needs, the main problem is the production of material goods.

Scientific research can affect this production in two ways:

(1) by reduction of general production costs, e.g. by decreasing the required capital supply of industrial investments, increasing productivity and improving designs.

The coefficient of reduction of such production costs is:

$$\eta = \frac{\Sigma S_t}{\Sigma S_s}$$

where ΣS_t represents the total costs per unit produced using traditional methods, and ΣS_s the total costs per unit produced by employing improved methods.

(2) by increasing potential capacity to meet needs. In some cases production is limited not by costs, but by a too low potential capacity caused by, for example, a shortage of raw materials or cultivated land. The results of research may increase these potential capacities by, for example, the use of substitute materials or more intense cultivation of the land. This effect can be determined by the coefficient of increased degree of covering needs

$$\varepsilon = \frac{D_s}{D_t}$$

where D_t represents the covering of needs (percentage) with traditional methods, and D_s the covering of needs (percentage) with employing improved methods.

The above two coefficients have different meanings in the process of meeting needs. Costs are a relative constraint which may be reduced by, for example, contracting loans. On the other hand a low potential productive capacity is a barrier which cannot be eliminated without the introduction of essential changes into the production process based on scientific research.

For this reason, when carrying out research, forecasting attention should be focused on the second aspect. This is demonstrated in a spectacular way by the need for research on alternative energy sources.

A number of reports concerned with the Unesco programme "Research and human needs" describe particular scientific research projects which can be utilized to meet the present needs. However, from the point of view of forecasting these projects feature some shortcomings; namely:

(a) they provide static statements without predicting the time of implementation, but such an approach fails to characterize a dynamic process;
(b) they are purely qualitative in nature and make no attempt to quantitatively assess the consequences which may follow as a result of the application of the research findings-
(c) they determine neither the place nor the time for fundamental research in innovation processes.

It is desirable that further forecasting studies be capable of answering these questions in relation, at least, to the research fields which are most promising in satisfying human needs.

In this discussion we are interested, first and foremost, in the role of fundamental research mentioned in (c).

FUNDAMENTAL RESEARCH FORECASTING

Even the formulation of this kind of forecasting causes a number of methodological problems. It is even more difficult to connect such forecasts with those concerned with future human needs. The related problems are as follows:

1. Possibility of Predicting Discoveries

Scientific discoveries, and especially those of a revolutionary character, can neither be exactly foreseen nor determined in time. Such a fact would seem to eliminate the possibility of forecasting fundamental research. However, as will be seen from a more thorough analysis, success in the matter is not impossible. Even if we are in full agreement with the Kuhn theory of scientific revolutions, we should bear in mind that old paradigms are replaced by new ones at intervals of scores of years or even more, but within these periods the evolutionary development of science can be predicted, if not in its particular discoveries at least in its general trends of progression. Today, a scientific discovery is very rarely the result of a single stroke of genius; usually it is a product of laborious research conducted by teams of specialists over many years. The researchers do now know, of course, the final results but they have formed an idea of the phenomena concerned with the expected results. If an inquiry is made (and this has been the case in France and the USA) in centres conducting fundamental research, it is possible to forecast fairly exactly the discoveries of an evolutionary nature (the revolutionary ones often being unexpected) for the subsequent 3-5 years, which is the usual period of fundamental research projects.

2. Generation of Applied Research by Fundamental Research

In this discussion we have started from human needs and proceeded to applied research intended for their satisfaction. The next step leads to fundamental research. Some difficulties are encountered here because the transition from fundamental to applied research is not direct. It can be imagined that fundamental research fills the reservoir of human knowledge from which applied research draws its inspiration. The system consists not of a single chain of events, but a network of interconnections; namely, a single scientific discovery may be applied in very diversified fields of applied research, while applied research uses knowledge from many different scientific disciplines. Nevertheless, there is almost always a major link in the network, and this is the scientific problem the solution of which will determine the further development of applied research (e.g. the explanation of the phenomena accompanying controlled thermonuclear reaction is indispensable for the subsequent state of the development of the peaceful utilization of nuclear energy). In the list of applied research topics oriented towards human needs, it is possible to ascribe to each item the fundamental research which will yield the results which are to be the starting point for the given applied research. From such an analysis a subsequent problem emerges.

3. Time Horizon of Fundamental Research Forecasting

It is estimated that the timescale between a scientific discovery and its full utilization in practice is about 7 years. This is a rough estimate because this period may vary within rather wide limits. However, although this period has been gradually shortened over the last quarter-century, the delay in obtaining practical results in relation to scientific discoveries is a rule having exceptions (for example, the transistor had been invented before the mechanism of semiconduction was fully explained).

INTERRELATIONS OF FORECASTINGS

From the above analysis of interrelations between applied and fundamental research it follows that most applied research (the practical use of which is expected in the following 5-year period), is based on fundamental research which has either been completed or is well advanced. Then, the forecasting of fundamental research related to human needs can concern only those projects the results of which are to be introduced into practice after 10-15 years. This conclusion also has methodological significance, because it means that the time horizon for needs forecasting must be longer than that for forecasting related fundamental research.

For this reason the long-term forecasting of human needs should be given more space in the programme "Research and human needs", and its results should determine the relevance of the research forecasting.

RESEARCH RELATED TO THE SPHERE OF NON-MATERIAL NEEDS

The above discussion was concerned mainly with research in exact, natural and technical sciences oriented towards material human needs.

In the field of non-material needs the role of scientific research is not so direct and the division into fundamental and applied research becomes at times impossible. The effect of scientific research on non-material needs can be reduced in a somewhat simplistic way to four facts; namely:

(1) the results of research are the basis for non-scientific decisions, such as political, social and economic;

(2) scientific research is helpful for the development of intellectual abilities and the instrumental aid to mental activity (computers, scientific and technical information services);

(3) conducting scientific research must go hand in hand with teaching activities at university level;

(4) the popularization of scientific research has an important, culturally creative function.

It follows from the above that the impact of scientific action mechanisms on the sphere of non-material needs is diversified and indirect. In contrast to material needs, in which priority goes to specialized research projects, here the importance lies rather with the general development of science, and mainly in the social sciences and humanities. Then, from the point of view of forecasting the importance is not so much in the content of the future research as in its quantitative range and the social role it can and should play. Here, questions of a more general nature arise: What form will scientific interest among widely varying social groups take in the future? How will the universities be capable of preserving their function as scientific centres with the ever-increasing teaching load?

LOCATION OF RESEARCH CENTRES

We have considered here the forecasting of the development of science on a global scale. New problems arise on the regional or national level, the most important being the distribution of the sources of scientific progress; that is, of research centres.

At the beginning of this working hypothesis we have assumed that the geographical distribution of centres of fundamental research does not necessarily coincide with that of the leading centres of technological advance. These centres need not be located in the regions where exist the human needs which they are trying to meet; but such a coincidence is desirable in many cases.

In preparing a forecasting exercise in more detail, it is necessary, then, to answer the question as to which regions should be chosen for the establishment of centres of fundamental research associated with human needs. This problem involves a number of contradictory factors. Reasons of economy and research efficiency dictate large, world-scale centres, while the political and social considerations give preference to the establishment of research centres in almost every country and certainly in each region. Location can also depend, of course, on the nature of research. Universal research in, for example, physics, chemistry and mathematics can be placed in a rather arbitrary way. But research in, for instance, geology or ethnography or on flora and fauna should almost exclusively be conducted in the specific region of the given subject of studies. Location of research centres is a vital element in the world's science policy. From the point of view of Unesco, the proposals on the location of research centres connected with human needs in developing countries are particularly important.

CONCLUSIONS

The forecasting of fundamental research oriented towards human needs is related to a number of methodological problems and is vital for the Unesco programme "Research and human needs". If this programme is to include real forecasting, and not only a static pattern of the present situation, this implies additional analyses and studies which should cover, among others, the following issues:

(1) definition of the fields of fundamental research determining progress in applied research oriented towards human needs;

(2) attempt at a quantitative estimation of the results of scientific research in meeting human needs;

(3) long-term forecasting of the evolution of the range and structure of human needs;

(4) analysis of intellectual human needs and their future development;

(5) forecasting of the geographical distribution of research centres,

(6) introduction of the forecasting of fundamental and applied research as an item in the Unesco programme "Research and human needs".

Characteristics of a useful forecasting effort: lessons from the analysis of past projects

D. L. Meadows

INTRODUCTION

The large number of highly placed participants at this meeting, and the careful preparations that have preceded the conference, indicate that several important programmes will gain impetus from our work here. In this paper I will draw on my own and on others' experience with scientific forecasting to offer several guidelines that might give those programmes more concrete impact on policy. Alexander King said at the opening of the meeting: "We hope that the usefulness of this meeting will be as great as the hospitality of our Georgian hosts." I will strive here to make that wish come true.

I have worked since 1970 on issues related to the topic of this meeting. A decade is not long compared with the professional experience that many of you bring to this conference; but two circumstances have provided me with many useful lessons during the past 10 years. The first was a book I wrote with Donella Meadows, *The limits to growth.*[1] About 3.5 million readers studied our book in one of 30 languages. Many of them took strong exception to our analysis and kindly offered to point out our mistakes. I estimate I have accumulated a pile of critical reviews that is over 3 metres high. Through study of the items in that pile I have gained some insights into the way thousands of people in many different cultures and different professions view scientific forecasting applied to human needs.

The second circumstance is my experience as director of a long-term forecasting institute. The Resource Policy Center at Dartmouth College is probably the largest academic institute in the United States devoted exclusively to teaching and research on the application of computerized forecasting models to long-term social issues. Through my involvement in work of the institute, I have learned many lessons about the factors that can cause forecasting studies to have (or fail to have) significant social impact. My remarks here will draw on the lessons gleaned both from my experience with the *Limits* and with the Center. I will divide my paper into five topic areas.

So that we all share some common impressions of our goals here, I will first describe some history of the Unesco programme on "Research and human needs"

(RHN), indicating the role this meeting plays in the larger programme. Second, I will point out two kinds of problems that have reduced the utility of earlier forecasting efforts. Next, I will suggest a new perspective on the role of forecasting efforts—one that points to the need for new guidelines for the conduct of useful work. Fourth, I will list 12 guidelines that we enforce on our own work within the Resource Policy Center. Finally, I will offer a specific proposal that could fulfil Unesco's desires to carry out long-range forecasting efforts useful in directing science and technology to the fulfilment of basic human needs.

HISTORY OF THE RHN PROGRAMME

The derivation of the RHN programme has been conveniently summarized in "Research and human needs", the report from the task force meeting held in Paris, 8-9 December 1980.[2] I paraphrase here from that document.

Meetings of experts in Venice[3] and Moscow[4] during 1975-6 gave birth to the idea of the RHN programme in the period when traditional development policies were starting to be reconsidered by agencies such as the World Bank, the International Labor Organization, the World Health Organization, and Unicef. Results during the first half of the 1970s Development Decade had demonstrated that economic development could not be achieved only by transferring money, technology, and expertise. Instead, the problems of development must be addressed in their appropriate sociocultural context. This point of view was reaffirmed in the paper by Dr Gvishiani, "Methodological problems of science and the evolution of human needs" in this volume. He pointed out that some earlier forecasting efforts failed when they focused on just one aspect of a problem in isolation from the many factors actually involved.

The RHN programme was proposed to Unesco at the same time as there began to be real concern for redefining conventional indices of development and for achieving a better application of the globe's scientific resources to the problems of development. The programme was given further impetus by those promoting the ideas of a New Economic Order.

The RHN programme can theoretically provide new directions for scientific activity, derived from a study of human needs rather than a pursuit of short-term corporate profit or military advantage (the primary stimuli to research today). But achieving this goal will require: (1) much greater elaboration of the relationship between basic research and the satisfaction of human needs, and (2) much more effective communication between researchers and decision-makers. Thus the Unesco programme has especially stressed development of new communication links, i.e. between researchers and those experiencing the problem, among researchers in different fields, and between researchers and policy-makers.

There are four main objectives for the RHN programme specified in the abovementioned task force reports:

(1) to mobilize and orient scientific research potential towards the satisfaction of human needs;

(2) to encourage increased interaction between decision-makers, scientists and practitioners in order to enable these groups to undertake a concerted effort in satisfying human needs;

(3) to promote research and pilot projects designed to demonstrate the

validity of the human needs approach; and

(4) to make available methodologies and provide information in this field so as to stimulate research and action programmes related to human needs and societal goals.

There has been significant growth in the programme since its inauguration in 1975-6. The first few scientists involved in the programme have been joined by many others to form a loose, world-wide network of personnel conducting a set of interrelated projects. Four groups—in Africa, Asia, Latin America and Europe—now carry out research on problems especially significant for their respective regions. A number of *ad hoc* groups conduct special projects with relevance to more than one region. And there are plans to expand the scope of the RHN programme even further.

Of special interest for this seminar is the new initiative that will involve forecasting studies. The Paris meeting concluded:

> Systematic and accurate forecasting of future events is more important today than ever before. Since relationships are more complex and changes occur more rapidly, it is vital to understand the potential consequences of present actions and incidents. It is recommended that Unesco, within the framework of the programme, Research and human needs, undertake forecasting studies on trends in scientific research that relate to the satisfaction of human needs.

Unesco has already initiated a series of books on trends in scientific research. But much is still to be done to achieve full implementation of the Paris meeting recommendation. This seminar in Tbilisi is an important step by those directing the RHN programme; through an examination of several important long-term forecasting efforts, it can help identify the most useful next steps in the effort to relate science and technology to human needs.

While much is already known about the prerequisites for a successful forecasting effort, Unesco must still proceed on its next efforts with caution; a careful look at past projects points to two potential problems.

TWO COMMON PROBLEMS

The first problem with most forecasting studies is nicely illustrated by a Persian folk tale involving, Nasrudin, a Mulla or Islamic teacher, who has lost his key.

> A man saw Nasrudin searching for something on the ground. "What have you lost, Mulla?" he asked.
> "My key," said the Mulla.
> So the man went down on his knees too, and they both looked for it.
> After a time, the other man asked: "Where exactly did you drop it?"
> "In my house."
> "Then why are you looking here?"
> "There is more light here than inside my own house."

When we examine past forecasting studies, we often find that they have focused their effort where there was light, not where the keys to important problems were to be found.

It is very tempting to study those issues and those aspects of issues best represented in available data. But yielding to that temptation automatically reduces the utility of the work. First, data are often available on those things that are easiest to measure, not those that are most important. Second, many data bases available today relate to issues that were considered to be important in the 1950s and the 1960s. Some of them remain important; but many of the most important factors influencing the satisfaction of human needs in the 1980s and the 1990s were not of critical concern before. Major defaults on international debt, declines in global oil production, and stagnation in agricultural output because of declining soil fertility are examples. Society has not developed major data bases on them; and forecasting efforts that are shaped to fit the data currently available will therefore be obsolete in important ways.

In the United States of America we often say that the army is always getting ready to fight the previous war. We must be careful in the work of Unesco that it is not always getting ready to study the important issues of the previous decade.

The second common problem with earlier forecasting efforts is of quite a different nature—most forecasting studies are carried out without any attention to the administrative mechanisms that would be required to implement their results. The problem goes beyond the need to make current institutions work as they were originally intended. We need entirely new administrative capacities to deal with the new kinds of problems that are likely to emerge over the next several decades. One example will illustrate this point.

Suppose for a moment that we had already developed a set of forecasting methods that were perfect—they could project with certainty all implications of the social and technical trends of interest to us. Suppose further that we had applied these techniques to the phenomenon of carbon dioxide emissions and concluded with certainty that sustained growth in the combustion of fossil fuels would lead inevitably to catastrophic changes in the weather during the early part of the next century. Would the availability of these "perfect" forecasting methods mean that we were able to deal with this profound threat to basic human needs? Not at all.

Indeed you can imagine the difficulties that would confront anyone attempting to respond to the results produced by the forecast. Many policy-makers would conclude with a sigh of relief that the problem would manifest only after they had left political office. Thus they would not see the need to do anything in response to the study. The poorer countries would immediately demand that the rich reduce their use of fossil fuels, so that some sort of national energy parity might be attained within the limits to fossil fuel use. Others would criticize the results by asserting that they overlooked the potential for some currently unanticipated new technology that will eliminate the problems caused by high CO_2 levels in the atmosphere. Many countries would probably accelerate their use of fossil fuels now in anticipation that future international action would be based on quotas tied to then prevailing levels of fuel use in each country. Still others would claim that the whole study was merely a ruse perpetuated by the rich countries to block industrial development in the poor.

My scenario might seem a bit far-fetched, but I can point to all these sorts of responses in that 3-meter stack of criticisms I have collected. And the CO_2 problem is relatively simple and straightforward. The problems cited above would be compounded in a forecasting study that involved human fertility, international trade, arms control, or aid programmes. Clearly, the advent of improved forecasting techniques is no guarantee that science can be used more effectively in satisfying human needs.

It is important to develop new forecasting methods. But without new approaches to the implementation of study results, enhanced forecasting capacities are likely only to heighten international strife and throw more intense light on the many forces that work to frustrate satisfaction of human needs. Of course I do not have detailed and final prescriptions that will lead to more effective implementation of results that emerge from Unesco's study of scientific trends and human needs. But the experience gleaned from the work of my institute over the past 10 years, and the conclusions derived from empirical studies of projects that have worked in the past, both point out one necessary perspective on the role of forecasting efforts that should be adopted by those who may manage Unesco's future efforts.

THE ROLE OF USEFUL FORECASTING STUDIES

Over the past 20 years there have been several examinations of the differences between those forecasting studies that were highly successful and those that were not. From the data they present, I conclude:

THE MOST INFLUENTIAL STUDIES WERE ONES THAT WERE MANAGED IN A FASHION THAT CREATED, LEGITIMATED, AND DISSEMINATED A GROUP OF EXPERTS ON THE TOPIC THEY ADDRESSED.

The projects' primary products were people, not reports. These people had became experts in the course of their study. They had become widely known and respected for their knowledge—legitimated through their work. And the project had provided a springboard from which they moved into decision-making positions in the institutions responsible for implementing the results of their project.

Successful forecasting efforts do require adequate data; they need powerful analytical techniques; they entail the preparation of formal reports and the development of formal models. But they also require sophisticated understanding of dynamics of organizational change and a commitment to empowering the analysts and advisors involved in the study. We need better data and improved analytical methods. But the current data and established methods are already far more advanced than our ability to respond to the results they provide. This can be illustrated with several examples already very familiar to everyone at this Seminar.

Our current forecasting techniques can already paint the rough outlines of the catastrophes inextricably attendant on the insane spiral of armaments currently under way among both rich and poor nations. Yet we seem incapable of achieving a weapons moratorium. It does not take massive new data bases to forecast the basic implications of the rapid increase in soil erosion that is in progress around the world in developed and undeveloped nations alike. Yet we seem incapable of implementing more effective conservation techniques. More sophisticated economic models are not required to identify the long-term damage done to those economies that are relying on rising foreign debt to finance the importation of oil. Yet we have not been able to find consensus on the nature of actions that can move the poorer countries off their petroleum dependence.

Urgently required in each case is a better communication between the scientists who have extrapolated current trends and the decision-makers who sustain them. This point was already acknowledged in the Unesco Paris task force report. What is new here is recognition that the principal way to achieve this communication is through the migration of personnel from research centres to decision-making agencies. Only when those with authority, resources, and responsibility for a problem in many different organizations begin to act in concert, with an internal consistency afforded by common goals and perspectives on the issue, will the many disparate responses to a major problem begin to sum to an effective programme. Time and again I have seen effective forecasting projects produce people with the necessary goals and perspectives and assist them in securing employment within the relevant action agencies. If it is to succeed, the Unesco programme on forecasting must place special emphasis on developing the procedures that can achieve that migration. In the third part of my paper I will indicate very briefly some of the techniques we have used to achieve it in our work.

GUIDELINES TO ACHIEVE EFFECTIVE COMMUNICATION

A useful forecasting project must be scientifically correct. It must employ appropriate data. It must use suitable statistical techniques. It must be very well documented. None of these goals is easy to achieve, and none of them guarantees that the project will be useful. Utility requires that many more guidelines be met by the conduct of the project.

The guidelines below, drawn from my personal experience, will be influenced to some extent by the fact that my forecasting efforts have always involved computer simulation models. Of course many of the projects potentially useful to Unesco will not employ computer simulation. Nevertheless, I believe our experience has rather general relevance, since it has been developed within a wide range of organizational settings and for a very broad set of problem areas.

For example, our group developed the model used by Presidents Carter and Reagan to project the long-term US and world consumption and price of oil, gas, and coal. We developed a model that will be used by 11 medical schools to teach doctors about the causes and cures for alcohol addiction. The US Forest Service has used one of our forecasting models to estimate the contribution that wood will make to residential heating in the next three decades. Congress has used another of our models to design improved legislative incentives for small-scale hydropower dams.[5-8] In each of these cases we used the following guidelines to ensure that our work would have an impact on the organizations actually responsible for effective change.

(1) Address the project to an issue that is of current concern to those responsible for action.

(2) Involve in the project from the start many members in the "invisible college", that group of people responsible for shaping public impressions and priorities on the issue.

(3) Conduct the project so that it produces very early (within months) a first running model and a set of preliminary conclusions. These should be disseminated widely to secure criticism that can be useful in focusing the final states of the project.

(4) Manage the project so that its quality, competence, and relevance are

widely perceived.

(5) Define partnerships with members of the invisible college; explicitly manage the project so that it serves the current personal and professional needs of those whose assistance you require.

(6) Select a prominent steering committee early, and educate them in the problem and the methodology of the project, so that they can be of material assistance to you.

(7) Embed the project in a teaching programme, so that the forecasting effort can become the means for raising many students to a level of expert knowledge about the problem and its solutions.

(8) Deliberately identify and cultivate the media contacts that you will need at the end of the project to disseminate the results.

(9) Expose your work to criticism often and thoroughly, but do not let misguided or uninformed reviews derail you from the central thrust of your work.

(10) Conduct your project in a way that reflects favourably on your students. Use the work to move them smoothly from research in your centre to positions of responsibility in the corporate or public organizations that actually control relevant aspects of the real world.

(11) When the basic methodological and data research are completed, arrange to transfer the project work with your students from your centre to the organizations they join. They should then work to adapt your forecasts and results to meet the day-by-day demands of the agencies that can respond to the research.

Each of these guidelines seems quite straightforward in the abstract. But they are difficult to implement in practice and they are ignored in the main by most forecasting studies. For example, I know that every one of you here works constantly within your own institution to identify your most outstanding analysts and to keep them, so that they might enhance the reputation, budget, and effectiveness of the work you do. This is just the opposite of my recommendation number 10. And when you have finally developed a mature and widely respected model, there is every incentive to keep it in house, so that it may be the basis for future work and subsequent funding requests. This is the exact opposite of what is called for in recommendation number 11.

If it is possible at all to design forecasting efforts that meet these guidelines, then it should be possible within Unesco. I have described above 11 attributes of a process that deliberately combines the conduct of scientific research with the achievement of education—exactly the two areas for which Unesco has been assigned responsibility. But even if you accept these guidelines, they cannot be carried out in isolation from real problems. The greatest damage to the utility of forecasting models is undoubtedly the notion that efforts to compile new data or develop new methods can succeed in isolation from a concrete, important problem and from a demanding, pragmatic client.

Many different problems are possible candidates for the focus of Unesco's next forecasting effort. I do not know what problems will ultimately be addressed by Unesco's new programme. But I do know several features that should be sought in the first few pilot projects. In the fifth and closing section of my remarks I will indicate those features and then describe a specific project that offers all of them.

A SAMPLE FORECASTING EFFORT

The problem addressed by the Unesco study should be important around the world but at the same time touch the life of each person. Here I find very useful a distinction proposed by Roger Levien. He has differentiated between global and universal problems. A global problem affects the whole world at once—for example growing levels of CO_2. A universal problem occurs around the world, but it takes a special form in each region—for example mounting soil erosion. No individual, organization, or region can have much sense of power over a global problem, for its solution lies only through transnational effort. However, the local manifestation of a universal problem is subject to local control. Hence there is much interest and responsibility for universal problems. I suggest that Unesco concentrate on universal problems.

If the first effort is to demonstrate quick and palpable success it should not deeply involve the factors governing economic and military competition regrettably manifest around the world. Problems related to the widening gap between rich and poor, and to the arms race, are of great importance. But the next Unesco project should practise first on something easier.

A third feature of the project should be its focus on issues with a large physical science component. Such issues more readily provide the base for significant cooperation across cultural, disciplinary, and political boundaries. Yet the problem should also require participation by biological and social scientists! The issue should be one that can be accommodated within the established programmes of major international institutes. It is unrealistic to expect that the austerity of the 1980s could give rise to a new institute just to facilitate the forecasting programme of Unesco, and it would be a waste not to involve the great resources that have already been brought together by IIASA, IFIAS, and their sister institutions.

Finally, the problem should offer the potential for a positive, constructive global vision even to those who are unwilling to accept utopian assumptions. It is crucially important to Unesco's RHN programme that it afford society some positive alternatives to the current trends, alternatives that give very concrete incentive for joint, and far-sighted, action. That will require a shared vision of an attractive future that seems realistically available.

Perhaps the first item on Unesco's agenda should be systematic identification of the several problems that satisfy the above criteria. I know that many participants in this Seminar could offer useful proposals. Simply to illustrate the kinds of problems that might emerge from a more systematic exercise, I will elaborate briefly on soil erosion as a possible candidate.

Deterioration in productive lands—agricultural fields, pastures, and forests—is rampant. It already touches the lives of the richest and the poorest on this globe. For example, the Global 2000 study projects that 40% of the world's tropical forests will disappear between now and 2000 if current trends continue. The US Department of Agriculture estimates that 40% of United States crop land is deteriorating at a rate that will produce serious declines in fertility in the foreseeable future. Yet new technology could be developed to arrest this deterioration at reasonable cost.

Different countries could cooperate on a programme in this area, exchanging forecasting methods, information, and experimental findings, without threatening their perceived economic and military interests. All countries would have something to contribute to the study, and none has yet achieved a fully satisfactory response to the problems of soil erosion. The vision of a land base that had been brought to a sustained and high level of productivity is one that could energize people and institutions at many levels.

NOTES

1. Meadows, D.H., Meadows, D.L., Randers, J. and Behrens, W.W. *The Limits to Growth*, Universe Books, New York, 1982.
2. Report on the meeting of experts, Paris, 8-9 December 1980, Unesco.
3. Report on meeting of experts, Venice, December 1975. Unesco Report SC.76/WS/25.
4. Report on meeting of experts, Moscow, July 1976, Unesco Report ED.77/WS/54.
5. Nail, R.F., *Managing the Energy Transition*, Ballinger, Boston, Mass., 1977.
6. Ramsey, B., Brown, P.W. and Kirshen, P.H. Proposals for changes in the taxation of hydroelectricity, RP 209, Resource Policy Center, Hanover, N.H., August 1979.
7. Marshall, A. and Norman, L. The dynamics of residential wood energy use in New England 1970-2000, RP 363, Resource Policy Center, Hanover, N.H., October 1981.
8. Goluke, U. A comprehensive theory of the pathogenesis of alcoholism. In: *The Biology of Alcoholism*, vol. 7. B. Kissin and H. Begleiter (eds), Plenum Press, New York, 1981.

Part II

Experiences of National and International Projects

The USSR Comprehensive Programme of Scientific and Technological Progress and its social implications for a 20-year period

V. Kotelnikov

The Comprehensive Programme of scientific and technological progress of the USSR for 20 years, and its social implications, was first approached in August 1972 within the framework of elaborating a long-range plan of national economic development for 1976-90. Its draft was approved by the 25th Congress of the CPSU. In line with the Congress' decisions the work on the Comprehensive Programme was continued and its horizon extended to the year 2000. In 1979 the project was completed. The conclusions and recommendations contained in the Programme were taken into account in the "Guidelines for the Economic and Social Development of the USSR for 1981-85 and the Period up to 1990" approved by the 26th Congress of the Communist Party of the Soviet Union. At present, in conformity with the government decree of 12 July 1979, the 20-year Comprehensive Programme of scientific and technological progress has become an original and integral stage of long-range planning in the USSR. The USSR Academy of Sciences, the State Committee for Science and Technology and the USSR State Committee for Construction will have to regularly revise the Comprehensive Programme (broken down into 5-year periods) and submit it to the Council of Ministers of the USSR and the USSR State Planning Committee (Gosplan) at least 2 years prior to the next 5-year plan period. The Programme precedes the draft Guidelines for the economic and social development of the USSR for 10 years elaborated by Gosplan on the basis of socioeconomic targets set by the Communist Party, and the Comprehensive Programme of scientific and technological advance. At present work is under way on drawing up a Comprehensive Programme for 1986-2005, which is of greater scale and more detailed than the previous one. It reflects the long-range perspectives of scientific, technological, and socioeconomic development of the USSR. It must diagnose the trends in social development, the origination of new demands, and capabilities of science and technology in coping with problems which arise; harmonize the efforts of all the echelons of science and production in accomplishing the long-term goals; identify the key national economic problems requiring priority concentration of efforts of scientists, resources and means; and provide the basis for drawing up large-scale goal-oriented programmes.

The Comprehensive Programme for 1986-2005 must scientifically validate the major directions of R&D of the scientific, technological and educational potential of the country; scientific and technological progress in the national economy with due account of domestic and foreign scientific and technological advance; and the socioeconomic premises and implications of introducing scientific and technological advance in the national economy, including the regional and international aspects of scientific and technological progress.

It should be noted that the Comprehensive Programme is not just a forecast aimed at predicting the development of uncontrolled events under one or other premise. The Programme offers recommendations to be taken into account in the process of plan elaboration, and forecasts the country's development following the accomplishments of the recommendations. The construction of such a forecast is obviously reasonable only in the event that the state is able to plan the national development.

The Comprehensive Programme was drawn up under the guidance of the Scientific Council of the USSR Academy of Sciences and the USSR State Committee for Science and Technology, and its commissions related to 27 branches of research (16 dealt with scientific and technological forecasts and the remaining 11 with socioeconomic problems). The Programme was developed by approximately 2000 scientists, specialists and practitioners representing over 500 research, and design and development organizations of ministerial and departmental reporting. The USSR State Planning Committee actively contributed to the Programme.

The Comprehensive Programme was based on hundreds of scientific and technological forecasts—economic analyses conducted by research organizations, teams of scientists and specialists. Following systematization and generalization, correction and specification in the commissions, the forecasts and analyses have acquired a comprehensive character. The commissions have also come up with specific recommendations, and have assessed the potential social and economic implications of their accomplishment. Simultaneously, the major socioeconomic problems were worked on. Taken into account were both the expected results of scientific and technological advance and the trends and goals of national socioeconomic development.

The Programme was not aimed at elaboration of an obligatory unified and optimal way of development. Different alternatives have been formulated. In some cases no selection was made; hence it contained several options. The choice would have to be made later, on the basis of more accurate calculations, social priorities and other considerations.

The Comprehensive Programme had to produce not only qualitative, but also approximately quantitative, recommendations providing for a balance between needs and capacities.

Coordination of the two aspects, i.e. scientific and technological and socioeconomic, was the most complex part of the activities related to the Comprehensive Programme elaboration. It was exercised through the special indicators, reflecting the social and economic results in each branch of scientific and technological progress, as well as on the basis of a comprehensive economic analysis. The work on the Comprehensive Programme helped identify the necessity of further research into individual scientific and technological problems which provided the basis for the subsequent formulation and development of separate goal-oriented scientific and technological programmes. For example, it is envisaged that a specific programme of manual labour mechanization will be drawn up providing for the development and manufacture of necessary machinery. Another programme concerns higher specialization in the engineering industry, and organization of inter-sectoral specialized production. Ferrous metallurgy will have to increase in its range and improve the characteristics of its end products. This is necessary for reducing the metal content in manufactured products, increasing the reliability of machinery produced, and cutting the costs of metal processing. The programmes should also cover the metal users to prepare them for employing more advanced products.

The Comprehensive Programme formulation revealed that the perspective energy supply will depend heavily on the possibility of reducing the specific fuel and energy consumption per unit of end-products and services. The engineering potential available in this area precludes the scale of energy savings that would allow for the reliable coordination of the energy complex and national economic development. Accordingly, a special programme was drawn up with a view to improving the energy consumption in all branches of production and non-production spheres.

The Comprehensive Programme indicates that the greater agriculture output can be achieved primarily through reducing transportation, storage, and processing losses as well as the application of genetics and selection achievements, i.e. biological factors. These recommendations will be further specified.

One of the prerequisites for automation and higher labour productivity is a more extensive application of electronic control and computing technologies.

A programme is needed that would rapidly boost the manufacture of computers and ensure their higher efficiency and more effective application.

In conformity with the Guidelines of scientific and technological progress the Comprehensive Programme sets priorities in individual branches of research. The priorities in science development form the starting point for the major directions of technological progress. Special emphasis must be placed on:

(1) Electronics (including computers) that must supply the national economy with new automation and data transmission facilities, and, in so doing, provide conditions for substantial increase in labour productivity.

(2) Power engineering, including projects in thermonuclear fusion, which in future will contribute heavily to the solution of the energy problem, and efforts in the immediate future relating to nuclear power stations, and the development of effective techniques for producing liquid fuel from coal. Unless the latter problem is solved, the growing scarcity of oil would hardly be compensated for.

(3) New technologies, such as laser processing, powder metallurgy, and the utilization of intensive catalysts.

(4) Agriculture: its mechanization, upgrading plants and animal breeds, and development of effective methods for microbiological protein production.

(5) Forecasting weather and climatic changes, including man-made changes.

(6) Study of the earth's structure, improvement of our knowledge on mineral deposits and fuel prospecting techniques.

(7) Medicine, especially in the area of widespread diseases.

Besides these, it is necessary to continue intensive research into living matter at cellular and molecular levels, on nuclear and elementary-particle physics, and space exploration.

In the humanities special emphasis must be given to the social and economic processes in modern society.

The rapid advance in these priority branches should contribute to the development of other fields of science.

The long-range plan of research, outlined in the Comprehensive Programme, is coordinated with the dynamics and scale of expenses on science development. The Programme sets the scale of instrument-making industrial performance necessary to support the research. Without appropriate development of this instrument-making industry, the R&D expenditure will hardly be effective.

The Programme also considers the issues of labour force training. Taking into consideration the duration of training cycles for scientists and experts it is already necessary to elaborate plans for training those scientists and specialists needed 10 years hence. There is much to be done in this area. The occupational retraining issue is a special case. Estimates indicate that according to the requirements of technological progress about three million people will change trades annually in the 1980s. It is necessary to retrain the personnel so that they can perform efficiently in their new jobs.

As previously mentioned, the Comprehensive Programme materials were utilized by Gosplan in formulating the Guidelines of social and economic development of the USSR and 5-year plans. It was rather beneficial and significantly improved the quality of the documents. When elaborating the Guidelines and 5-year plans, more specific and detailed estimates were carried out taking due account of the circumstances. The approximate figures of the Comprehensive Programme have been specified and changed but the general tendency was mostly retained.

The elaboration of the Comprehensive Programme of scientific and technological progress of the USSR for 20 years thus provides a bright illustration of the growing role of science in the life of our society.

Interfutures: an assessment of the project's development

J. Lesourne*†

INTRODUCTION

It was at the end of 1975 that the council of OECD decided to launch the Interfutures project to study "the future development of advanced industrial societies in harmony with that of developing countries". The council justified its decision by the following considerations:

> A number of new problems have emerged which have important implications both for the advanced industrial countries and with regard to their relationship with the developing countries. Difficulties in maintaining growth and full employment, supply and demand imbalances in markets of important commodities, inflation, unprecedented balance of payments problems, concern about the new patterns of trade, investment and monetary relations and the growing imbalance between the developed and the developing economies are typical examples of such problems.

Hence, it was asked to carry out a future-oriented analysis, identifying the key factors and their relationships, assessing likely future developments and emphasizing their implications for short- and mid-term policy options:

(a) an assessment of longer-term economic developments and relationships among advanced industrial societies;

(b) an assessment of longer-term developments in the relationships between the advanced industrial societies and the developing countries;

(c) an assessment of potential physical constraints and opportunities related to energy, raw materials or the environment on future development;

(d) an assessment of the future evolution of international economic structures and systems in response to growing interdependence.

---

*Director of the Interfutures Project (1976-9).

†The opinions expressed in this paper are strictly personal and do not necessarily represent the views of the OECD Secretariat.

The results of the project are well known, since they have been published in the final report.[1] Therefore, the purpose of this paper is neither to present the conclusions obtained nor to evaluate them. It is to describe the development of the project and on this basis to propose a critical appraisal of this development. Consequently the following aspects will be successively considered:

(1) the organizational framework,
(2) the methodology,
(3) the difficulties encountered,
(4) the lessons of experience,
(5) the contributions to long-term analysis development, and
(6) the relevance of long-term research to policy-making.

THE ORGANIZATIONAL FRAMEWORK

The project started in the middle of January 1976 and was supposed to end on 31 December, 1978. In fact, in order to disseminate the results and to prepare the publication of the report, a reduced team of six scientists remained until 31 October, 1979. In evaluating the length of the project, it should be considered that, at the start, only the Director was present and that offices had to be found and equipped.

The total budget available for the 3-year period amounted to approximately 4 million dollars. To this have to be added contributions from three Foundations (Toyota Foundation, Ford Foundation, German Marshall Fund) which financed special tasks within the project and a few hundred thousand dollars of government's contribution for 1980. These amounts covered the full cost of the project including rent and equipment of offices, furniture, staff expenditures and participation to OECD Secretariat overheads. Hence, they must be compared to the expenditures of a corporation and not to the cost of a project as usually estimated within a University.

To develop the project, a special structure was created within the OECD Secretariat. It was composed of: an independent Directorate with a full-time team and consultants, a council, and an Advisory Board.

The Directorate

At its maximum size the full-time team was composed of 15 scientists (including the Director and three Deputy Directors) and 10 other staff members. Among the scientists, only one-third stayed for 3 years, the others having contracts for a period of 1-2 years. A good international representation was achieved since the scientists came from 10 different OECD countries. Their training was in systems analysis, macroeconomics, engineering, sociology, political science, and demography. They had experience of civil service, research and education or business.

In addition, the team was able to secure the services of approximately 60 consultants for periods of between a few days and 1 year. A special mention must be made of the 15 consultants from developing countries, since their contribution to the project has been extremely important.[3]

The Council

The task of the Council was to supervise the development of the project and to approve the main orientations. Each of the supporting OECD governments was represented at the Council by one delegate chosen by it, and who was a high level civil servant. But, due to the heterogeneity in government organizations, the department to whom this civil servant belonged differed widely from one country to another (Treasury, Budget, Economic Affairs, Planning Agency, Central Policy Staff, etc.). The Council used to meet two or three times a year under the chairmanship of the Japanese representative (Dr Okita and later on, Dr Miyazaki from the Planning Agency). For budgetary matters, the council's approval was only a first step in a long procedure finalized at the level of the OECD Council.

The Advisory Board

Composed of 10 leading personalities (Professors Allardt, Aron, Bell, Fels, Oshima, Raynaud, Ruffolo, Seers, de Wolff, and Dr Gyllenhammar), the Advisory Board was appointed by the OECD Secretary General on the proposal of the Director. Its purpose was to give advice to the Council, or to the Director, on the development of the project. It usually met twice a year for 2 days each occasion. The members were also individually available for discussions with the team on special issues.

The Advisory Board has proved to be a very useful body during the whole operation.

The Directorate had not only to give life to this structure, specific to Interfutures, but also to establish and maintain contacts with the other Directorates of the OECD (including the Development Centre) and to attend various committees of the Organization. Such relations were essential to have access to the information existing within the Secretariat and to understand better the current policy issues.

This organizational framework was basically sound and probably much more satisfactory than for many similar projects. It generated, nevertheless, a few difficulties which will be discussed later.

THE METHODOLOGY

The decision of the Council did not impose any research programme and the first task of the Director was to prepare with the Deputy Directors such a programme and to present it at the first meeting of the Council.

This programme was based on a small number of key options. The project was conceived as:

(1) an effort in prospective analysis (prospective analysis tries to describe the different futures that could result from the behaviour of the "actors" involved within the limits of the systems through which they act);

(2) an attempt to perceive interdependence without losing sight of specific problems;

(3) an approach which was primarily economic but took into account the sociopolitical context;

(4) research concentrated on the problems of relevance to OECD governments.

Such a choice implied two important consequences:

(1) It was decided that the team should not build a new world model, since this task would consume a major share of the resources and would have to be launched before a real understanding of the issues at stake.

(2) The programme was organized to combine partial quantitative analysis, qualitative studies and explorations with global models.

Hence, after the 6 months preliminary phase of programme definition (phase 0), the project was divided into four phases:

Two preliminary phases (phases A and B) These were devoted to an analysis of the literature in order to assess the facts already known, the main relations between fields, the principal uncertainties, and the key policy issues. These two phases coincided with the creation of the team and lasted approximately 1 year in total. Their result was the edition of 21 documents (between 50 and 100 pages each) devoted to the main fields of analysis (international relations, population, energy, food and agriculture, monetary issues, internal problems of advanced industrial societies, development strategies in the Third World, elements for scenario-building, just to give a few examples). These documents were sent to the government to initiate a dialogue with their specialists.

The third phase (phase C) This started immediately after, and lasted till the end of the project. The research was then concentrated on a small number of topics which obviously needed in-depth analysis. These topics, which had been identified at the end of the preliminary phases, were the following:

(1) the prospects for value changes within advanced industrial societies,
(2) the macroeconomic perspectives of advanced industrial societies,
(3) the dynamics of structural change in advanced industrial societies,
(4) the long-term development issues for the main countries and regions of the Third World,
(5) the sectoral redeployment of world industry.

Simultaneously, a set of scenarios was progressively selected to illustrate possible paths of world evolution. The scenarios tried to embed assumptions along a few key dimensions concerning:

(1) the relations between OECD developed countries,
(2) the internal dynamics of the developed societies,
(3) the trend in relative productivities between OECD countries,
(4) North-South relations and the relations between LDCs.

While, during the preliminary phase, the team structure was loose, each member having its own task, subgroups of team members and consultants were created during phase C to take in charge the research areas which have just been mentioned.

Though the scenarios through their very definition were more than a sheer model run on a computer, they needed a central core of consistent data. In order to do so, the first idea was to use in parallel several global models and to compare their results for roughly similar inputs, but it quickly proved to be unfeasible, less for technical reasons than for accessibility.

Hence, the team decided to use SARUM as its central model; such a solution had decisive advantages:

(1) One of the members of the team (David Norse) had contributed to the building of this model and hence knew it thoroughly.

(2) SARUM was the property of one of the supporting governments and hence was easily put at Interfutures' disposal.

(3) The British administration kindly accepted to introduce into SARUM the main changes asked for by Interfutures (for instance, an increase in the number of regions), and that at a reasonable cost.

(4) SARUM had two features in which the team was especially interested: the possibility of introducing trade barriers through price distortions, and the existence of possibilities of substitution at production and consumption levels.

A large access to data was necessary for phases A, B and C. The team had at its disposal the OECD data bank for member countries, the OECD Development Center Statistics for developing countries. In addition, it constantly used the services of DRI (Data Resources Inc.) through a terminal installed in its offices. The World Bank kindly agreed to put its figures at the disposal of the team through the DRI connection. All these statistical sources were nevertheless insufficient for the prospective analysis of given industrial sectors. In that case the team or the consultants had to look for unpublished professional sources.

The last phase (phase D) This started during phase C and finished at the end of the project. It was the diffusion phase. The main activities corresponding to this phase were:

(1) the organization of a 2-day meeting of high-level civil servants representing the member countries (this meeting discussed a "policy issues" paper prepared by the team in cooperation with the Council),

(2) several visits to member countries to present and discuss the results to the most concerned government departments,

(3) a final report published in 1979 under the OECD Secretary General's responsibility,

(4) a summary of the final report for wider circulation inside the governments,

(5) a special issue of the OECD Observer,

(6) a programme of conferences to present the outcome of the project to various audiences in many countries.

To have a clear view of the project follow-up is difficult because of the diversity of initiatives taken at countries levels. Just to give some examples:

(1) The report has been translated into Japanese by the Japan Productivity Center and into Spanish by the Spanish Government. Partial translations exist in several other languages (German, Dutch, Greek, Finnish, etc.). The report has been one of the basic documents for the French 8th Plan; the Belgian Government has organized a three-day conference to discuss it; Australia and New-Zealand have used it as a starting base for a more comprehensive analysis of their specific problems.

(2) The report has been presented to the main OECD Committees (Economic Policy, Industry, Agriculture, Environment, ...) and several committes have included on their program a study of the problems raised in the report.

This brief description of the methodology should be sufficient for the reader to understand the difficulties encountered and the lessons drawn from the experience. For a more detailed description see Note 2.

THE DIFFICULTIES ENCOUNTERED

These may be grouped under four headings: time, team, communications with governments, policy analysis.

Time

Most of the team members joined between March and October 1976. They started to leave during the first 6 months of 1978. Hence, the team was really fully operational when it began to disband. Such a situation had several disadvantages:

(1) Many activities had to be followed simultaneously during the second half of the project, for instance scenario-building and report-writing!

(2) It was not possible to develop the scenarios as initially planned (with the elaboration of subscenarios for each of the main industrial sectors). For the same reason, the special task of analyzing discontinuity scenarios had to be dropped.

(3) Many good internal working papers (though their content was used for the final report) could not be brought to final completion and circulated to governments.

(4) The evaluation of policy issues had to be much more limited than would have been ideal.

On the whole, it would have been preferable, within the same budget constraint, to develop the project over 5 years. Since the "take-off" and "landing" periods would not have been longer, the "cruise flight" duration would have been increased by at least one-third. The experience of IIASA's big projects seems to confirm this observation.

Team

Due to OECD reputation, the recruitment of the team members and of the consultants was extremely easy, and on the whole the quality of the individuals involved in the project has been outstanding. The cultural diversity, both nationally and professionally, was one of the great assets of the team. But this tremendous advantage had nevertheless its negative counterparts:

(1) Many members of the team had to move their family and to organize their family life in Paris for a small period of time. Hence, significant "transition" costs in money and time had to be accepted.

(2) Twenty to twenty-five individuals who come from 10 different countries, and who have never worked together before, do not create a team in one day; the linguistic difficulties and the cultural barriers must not also

be underestimated.

(3) Prospective analysis (or systems analysis) is not a discipline like econometrics or mathematics which is taught on the same basis in every country. Therefore, the members of the team had not spontaneously the same views about the purpose of the project and the way of pursuing it. Consequently, as the Director of the project, I adopted a method different from the one I would have selected with a national team. Instead of immediately imposing a debate on the precise methodology and on the role of scenarios and models (such a debate could have generated hard and artificial conflicts), I preferred to obtain an agreement on only the broad lines of the research programme and to start with the discovery of the issues (phases A and B). Hence, the methodological consensus was reached naturally, step by step, as the project developed.

In spite of the staff quality, the personal achievements of the individual consultants or full-time members differed widely. Unfortunately, the time constraints made the deficiencies of some analysis difficult to complete or improve as much as I would have wanted. Here again, a longer duration would have given a better control of the operation. To give an example, I consider that the prospective analysis of the International Monetary System should have been more comprehensive.

The fact that we have not been able to use several models in series or in parallel is also regrettable. The confrontation of several models in parallel puts into evidence convergences and discrepancies and improves judgment; the use of a more specific model within the framework of a more general one enables a more precise understanding of important issues.

Communications with Governments

Communications with governments are (relatively) easy when centred on precise issues followed by a given unit in the governmental structure. Needless to say, that could not be the case for the Interfutures project. Many governments had no special units for long-term analysis and even when such units did exist they were frequently cut off from policy-analysis mechanisms. Hence, the governments selected, to represent them on the Council, civil servants from a great variety of departments. These civil servants were in charge of only a few of the issues raised by Interfutures. Often they found it difficult to create a national committee or network to follow Interfutures. As a consequence, the quality of the communication was extremely different from one country to another. The best results were obtained when an Interfutures team member of a given nationality, having an intensive knowledge of his own administration, could cooperate with the corresponding Council Member to organize at the national level the follow-up of the project.

The documents of preliminary phases A and B had been written to stimulate discussions with governments and especially to identify areas of agreements and disagreements in the main fields covered by the project, but the comments received came frequently from persons having insufficient knowledge of the purpose of the project and who reacted to the text of a paper without understanding its aim. Thereafter, it was decided to circulate only much more elaborated papers, at the expense of the intensity of exchanges.

The communications were easier for the final report and the agreement to publish it was obtained without substantial difficulties.

There remains nevertheless the fact that, for partly superficial and partly fundamental reasons, different governments were interested in different aspects of the Interfutures project.

Policy Analysis

After the completion of the final report—but perhaps before its diffusion—it would have been useful to select a few precise policy issues to evaluate possible consequences of alternative courses of action. But the time constraints did not permit it; not would it have been feasible earlier. There is a necessary sequence in the development of a global project like Interfutures: on one side, preliminary phases play a major role in the final outcome since they enable the team to understand the world issues before embedding model runs into scenarios; on the other side, the synthesis has to be sufficiently advanced for the evaluation of precise policy alternatives to be done within a global framework.

THE LESSONS OF EXPERIENCE

If the Interfutures project had to be done again, but with the experience of the preceding exercise, which modifications would I suggest for the development of the project? On the assumption that project duration could not be lengthened, I would essentially introduce two changes:

(1) Round tables with representatives of member countries would be organized on each of the main fields identified for the preliminary phases (which would mean approximately a dozen round tables. A short discussion paper would be prepared for each. Such a procedure would allow one to discover more quickly the information available, the key issues and the areas of uncertainty. It would also increase the team communication network in each country.

(2) A greater effort would be made to obtain, in each government, the creation of a special committee on which most of the interested departments would be represented. The names and positions of these committee members should be known so that direct communication could be established with them, the Council Member being, of course, regularly informed.

THE CONTRIBUTIONS TO LONG-TERM ANALYSIS DEVELOPMENT

Considering now long-term analysis in general, I would like to raise five important issues:

(1) Most of the global projects have been developed on too short a period. The optimal length is probably around 5-6 years. Such a timescale is necessary to define the research programme, to constitute a team (a team is more than a collection of individuals), to understand the problems at stake, to have a good knowledge of the existing information, to build adequate models, to conceive scenarios which are meaningful, to study precise policy issues, to eliminate deficiencies at the various steps, and to produce a broad variety of final documents. A very small group is then necessary for a couple of years for the systematic dissemination of the results.

As for the size of the group, it may depend on the very nature of the project but the existence of a core of full-time people following the project from beginning to end is an absolute necessity. This core must have a minimal size—five to six scientists for instance.

(2) A global project should not be reduced to the building or to the use of a model. It is a much broader exercise including a preliminary identification of the system to be considered, an analysis of the possible issues, the elaboration of data, a combination of various approaches (qualitative studies, quantitative studies, global model runs) in order to exert judgment on possible trends or policy outcomes, etc. Simple models may have to be used at different steps of the process. If, nevertheless, a central model is built: it should not be at the start, and the model should progressively develop out of simple models as the understanding itself develops.

Rather than one model, a family of models is preferable, with either (as in IIASA food and agriculture project) a model for each of the key countries and an interaction model linking them or (as in other exercises) a global model coupled with more specific models on sectoral domains.

(3) Prospective analysis should try to identify the agents whose behaviour is able to influence the future evolution (governments, multi-national corporations, ethnic minorities, social groups, trade unions, etc.) and to describe the goals aimed at by these agents, the expectations they have, the constraints defining their choice sets, the policies at their disposal. It is easier then to localize potential conflicts and to question their possible outcomes. The danger of most global models is that they eliminate social agents, replacing them by abstract entities like regions or continents led by fictitious governments. It reinforces the risk of considering governments as almightly decision-makers who can select policies freely. It is important to introduce game structures into global analysis and to accept players other than governments.

For the same reasons it is dangerous to adopt a purely economic view of the world. In the long term, economic, social, political, ecological aspects are necessarily intermingled. For instance, many proposals concerning North-South relations are unrealistic because they do not take into account major political issues.

Hence, to limit a scenario to a model run is highly unsatisfactory since a scenario should be defined by much more basic assumptions about the consequences of the tensions existing within the system under study.

(4) In the evaluation of the cost of a global project, the time and money which should be spent on external communications during and after the completion of the project is always grossly underestimated. Though I have not made precise computations, I estimate its optimal amount to be at least 50% of the total budget! For most of the global projects, including of course Interfutures, these efforts have been too limited.

(5) A global project is always influenced by the nature of the institution which hosts it and by the composition of the team which develops it. There is no such thing as the unique objective answer to given terms of reference. Hence, it would be important to have several projects going on simultaneously throughout the world. They could differ by the supporting organization (UN, IIASA, COMECON, OECD, Group of 77, a national agency, a university group, a private research centre, to give a few examples), by the main emphasis, by the key technical (or political) options of the team. The future is nobody's exclusive property.

As we have explained, each of these projects should last long enough, and receive enough funds, to produce a significant contribution to the understanding of the long-term issues. Exchanges of views and regular communications between the permanent teams in charge of these projects would certainly improve the formulation of important multilateral policy problems. Conflicts would not disappear, but the approach to compromises could be conceived in a broader context and substantially ameliorated.

CONCLUSION

Everyone is now convinced that important problems cannot be settled by a Seminar of Heads of State. Decision-makers are everywhere and nowhere! More precisely, political leadership is essential but it cannot do much without the support of civil servants, political parties, social groups, public opinions, and so on. Therefore, it is unrealistic to believe that a global project will immediately have a decisive impact on practical decisions. We can even predict that the more global it is, the less apparently effective it will be. These trivial remarks suggest several courses of action:

(1) A first possibility is to associate to the project one or two precise proposals and to study them deeply enough to be reasonably sure of their feasibility. The follow-up action should then be concentrated on these proposals.

(2) Another solution is to try to modify the perception of world problems which various social groups live on. Then, the impact of the project is only indirect and practically invisible, but it does not mean that efficiency is smaller.

In that case, the message should take the form of different outputs: a full-fledged scientific report, executives' report, papers for daily press and periodicals, TV shows, widely circulated and easily accessible books, exhibitions, lectures, seminars and conferences. But this part of the job should not be totally left in the hands of scientists. Communication specialists should be brought in to identify the targets, to evaluate their initial knowledge, to describe their expectations, to define the appropriate messages, and to decide about the appropriate media or supports.

I do not share the naive view that science can by itself solve the difficult problems of the world, nor that governments of scientists would be definitely better than existing ones, but I am convinced that a scientifically oriented analysis of global problems may serve to improve man's future.

NOTES

1. *Facing the Future: mastering the probable and managing the unpredictable.* OECD, Paris, September 1979.

2. Lesourne, J. and Malkin, D. The Interfuture study: some reflections on methodology. *Futuribles*, August-September 1979.

3. Of course, they presented their own independent analysis of developing countries' problems, the final report being only issued under the responsibility of the Director.

The Interfutures study: some reflections on methodology*

J. Lesourne and D. Malkin

INTRODUCTION

Prospective analysis—the investigation of possible futures as a prerequisite for drawing up strategies or policies for action in the present—took on a new lease of life in the 1960s. Its development was favoured by an awareness that traditional analytical or forecasting methods only partly or unsatisfactorily met the need to master the structural changes which are a feature of developments in the world today, and did not take sufficient account of the growing interdependences within economic and social systems, both national and international.

As everyone knows, however, the emergence of any issue—energy, land-use planning, international economic relations, the balance of strategic power—as a subject of prospective analysis is not only dated but also socially and institutionally situated in so far as it involves certain stakes, conflicts or potential upheavals—or when it is so perceived by those who feel that prospective analysis is necessary in that area.

The Interfutures exercise is of interest from both these standpoints: the date of its launching, 1 January 1976, i.e. in the midst of the last recession; and the place—the OECD, intergovernmental organization whose membership comprises the developed market-economy countries.

These circumstances, and the research team members' own experience, led to the adoption of a prospective approach which this article will seek to describe.

From the terms of reference it can be seen that Interfutures had three characteristics which, from the outset, distinguished it from most other OECD research projects and many other future-oriented studies;

First of all, it was explicitly stated that the aim was to analyse long-term problems in a global context, stressing the interdependences both

---

*This text is an abridged version of a paper published in French in *Futuribles*, 26/27, September-October 1979.

between the sectoral aspects and between the strategies of the actors involved (mainly States or groups of States). Then (but this was already more implicit) it was realized that the time horizon chosen implied that the study must necessarily go beyond economic analysis as such and take account of political, social and cultural factors. Lastly, the project had to analyse simultaneously the internal problems of the developed societies and the relations between them, as well as the interactions between them and the developing societies.

Another characteristic of Interfutures is one which sets it apart from more academic "prospective" studies which tend to be concerned with the future *per se*: Interfutures' end purpose was to examine the policy choices open to specific actors—usually the governments of the developed countries.

The methodological reflections which follow are arranged into three sections: Section I deals with the difficulties which are inherent in prospective analysis as applied to international relations, and which gave rise to the questions the Interfutures team had to ask itself after making a critical survey of the world future-oriented studies available at the beginning of 1976. Section II, which is, of course, the heart of this article—presents the methodology Interfutures worked out. Here prominence will be given to the role of the scenarios and the stages which governed their construction. Section III, in conclusion, will attempt to make a critical assessment of the method used, and to show what can be learned from the Interfutures study for the future development of prospective analysis.

I. INTERFUTURES AND PROSPECTIVE STUDIES ON INTERNATIONAL RELATIONS

The methodological problems faced by Interfutures cannot be properly analysed without making a critical survey of some earlier exercises in future-oriented world studies, or without any reference to the specific difficulties inherent in the prospective analysis of international relations.

Past Experience and its Limitations

Prospective studies carried out directly and explicitly on a world scale began to develop in the 1970s. It was only proper that Interfutures should take account of at least some of the considerable experience acquired—both the information which had been gathered and the methodologies used.

Previous studies can be classified according to the following criteria:

(1) whether or not they used modelling,
(2) whether they were global or sectoral,
(3) whether or not they took account of geopolitical factors and actors' strategies in a historical perspective.

Table 1 shows some of the most significant prospective studies taken into account by Interfutures, arranged according to this classification.

(a) The emergence of long-term models, their advantages and limitations. Global modelling had its origins in the conjunction of two phenomena, and its successful diffusion and impact are, in some ways, attributable to this fact.

TABLE 1 Classification of some World Prospective Studies

		Modelling used	Modelling not used
Global	Actors' strategies		
	Yes	IIASA "Agriculture and Food" project [11]	RIO [6]
	No	Meadows [1] Mesarovic-Pestel [2] Leontief [3] Bariloche [4] SARUM [5]	
Sectoral	Actors' strategies		
	Yes		GRESI [8] WAES [9] Sectoral studies UNIDO [10]
	No	MOIRA [7]	

[1] D. Meadows *et al.*, *The limits to growth*, MIT Press (research carried out for the Club of Rome).

[2] M. Mesarovic and E. Pestel, *Multilevel computer model of world development system*, IIASA, Laxenburg, Austria, 1974.

[3] W. Leontief *et al.*, *The Future of the World Economy*, United Nations, 1976.

[4] A. Herrera, *Latin American world model*, Fundacion Bariloche, Buenos Aires, 1974.

[5] Systems Analysis Research Unit Model, Research Report of the British Department of the Environment, London, 1977.

[6] J. Tinbergen *et al.*, *Reshaping the International Order*, Report to the Club of Rome, Dutton, New York, 1976.

[7] H. Linnemann, *MOIRA-A model of international relations in agriculture*, North Holland, Amsterdam, 1976.

[8] GRESI, *La Division internationale du travail, Trois scénarios contrastés*, Documentation française, Paris, 1976.

[9] C. Wilson *et al.*, *WAES—World Alternative Energy Strategies*, MIT Press, 1978.

[10] UNIDO Centre for International Studies (some of these are founded on actors' strategy hypotheses; others not).

[11] International Institute for Applied Systems Analysis, Laxenburg, Austria. "Food and agriculture" project (launched the same year as Interfutures and still continuing).

The first of these phenomena is awareness of the increasing interdependence between national economies, between the various aspects of world economic development (resources, production system, consumer habits, etc.), and between the economic and the social sphere in the widest sense of the term.

Global models, especially those which have recourse to systems analysis, are particularly useful tools for handling the different aspects of this interdependence dynamically by coupling formalization and data processing.

The second phenomenon, more subtle and also more dangerous because often concealed, is the one which tends to substantiate the argument of the inadequacy of economic theories—or, more exactly, of those theories which seek to explain socioeconomic developments at international level. This has led to the resurgence of a certain temptation to empiricism, which tends to creep into modelling exercises, even though some of them may try to hide it behind a veil of theory. Thus the desire to be all-encompassing, which results from the taking into account of interdependences added to the mass of data to be processed quantitatively, means that statistical estimation takes precedence over the reasoning of political economy and over historical perspective.

Certainly the catalyst role which models have played—and are still playing—should be emphasized. Whatever the worth of their conclusions, they have the not inconsiderable merit of drawing the attention and opening the eyes of a wider public to a number of fundamental questions such as raw materials supplies, inequalities of income in the world, and geographical distribution of economic activities. Collectively, they have certainly helped to present a number of key issues in a global perspective and so influenced policy formulae in national and international institutions. This, however, is a merit which must be regarded with circumspection when their results are largely invalidated by more sectoral studies or when the formalization of international economic relations leads to analysis being developed in a geopolitical vacuum. *It is these two drifts, especially, that Interfutures has striven to master in its use of a global model.*

The decisive advantage of models lies, no doubt, in the fact that they provide a consistent accounting framework for the variables whose evolution they seek to determine. Whether the model be normative (Bariloche, one of the Leontief model variants) or exploratory (Meadows, Mesarovic-Pestel, SARUM), accounting consistency, as compared to the juxtaposition of extrapolations or sectoral prospective analyses, offers two advantages. First, it brings out the interdependences between sub-systems whose dynamics are linked. Second, it may permit the identification of possible inconsistencies or major imbalances which become apparent behind the constraint of formal consistency. Unfortunately, as will be seen, only rarely is this second advantage of modelling turned to account, in so far as it leads to questioning the stability of the structures on which the models rest.

Models have another advantage: being quantified, they make it possible to obtain orders of magnitude. However, some prudence is still required: the models' stability may be only the apparently positive side of a structure which is resilient but not relevant. Furthermore, effective use of orders of magnitude given by models certainly demands that they should not be confused with projections. We have to be able to measure their reliability, in particular by evaluating their dependence on the basic hypotheses and by comparing them with sectoral prospective studies. In making use of a world model, therefore, Interfutures took care that orders of magnitude concerning world distribution of industrial production were not only explicitly linked to the hypotheses of the scenarios, these hypotheses being expressed quantitatively for introduction into the model, but were also evaluated in the light of parallel sectoral studies.

Lastly, a further benefit of modelling is that it makes it possible, while using a consistent global framework, to spotlight certain specific issues. This is the advantage of sectoral, or predominantly sectoral models. Obviously, this technical possibility is useful only if the relationships between the disaggregated sub-system and the rest of the system are relevant.

Models do have some essential limitations, due to the nature of the structures they imply. Several questions have to be mentioned here: even if, after statistical estimations, a global model can claim some degree of reliability in representing a system's past trends, it still cannot be concluded that the same representational structure must necessarily be used in the future. A fundamental characteristic of possible developments in the world economy lies in the fact that structural changes can equally well occur in the long term within developed and developing countries, or between developed countries, or between them and the developing countries. It is, however, far from easy to identify such structural changes when using a method which postulates at one and the same time the invariance of a system of relationships between countries or groups of countries and the homogeneity (not to say the oneness) of the logics of development linking the macro- and microeconomic aspects. Because of the very nature of the models based on such a hypothesis, methods of this kind lead either to a schema of gradual changes caused by the generation within the system of re-equilibrating forces (generally incorporated in the model), or to a disequilibrium of the final state which may indicate the main instabilities the system will encounter in its development but will not make known the concrete forces at the origin of this disequilibrium or these instabilities.

These general limitations make it necessary to supplement the use of models by more qualitative approaches which enable us to give a more explicit account of possible structural changes:

(1) Approaches whose point of departure is a reflection on the qualitative or quantitative causes which may modify the balance of power represented in the models. The models neglect the important role of the various actors whose past behaviour and strategies have given rise to the types of relationships formalized. It seems that at global level only *historical perspective*, by throwing light on the evolution of the aims and the political, economic and social strategies open to the actors, as well as on the possible conflicts between them, can help identify structural changes.

(2) Approaches devoted to partial prospective analyses whose refinement (like their limitations) is due to the fact that they are not subject to the constraint of global consistency. The aim of such analyses is to reveal the possible structural changes masked by this consistency constraint.

(b) Prospective analysis without the use of models. Its advantages and disadvantages are apparent by contrast with what has been said about models. We shall thus be fairly brief—but perhaps more concrete—on this score.

Two of the non-modelled world future-oriented studies available to Interfutures at the outset or in the course of its research seek to take explicit account of geopolitical factors in historical perspective: the Club of Rome's RIO project, and the GRESI scenarios of international division of labour. The others, such as WAES, which dealt with the prospective study of energy, or certain UNIDO world sectoral prospective exercises, while more traditional in conception, are based on the strategies and policies of the various actors involved in order to draw up long-term supply and demand scenarios.

If we take as an example the RIO project, one of this project's main features is that it attempts to take explicit account of the demands for a new world economic order expressed by the Group of 77 in the international organizations. It is thus both political and economic in perspective, and possible (or desirable) developments in the international environment are tackled explicitly in terms of strategies in case of conflict or cooperation between actors. RIO's work on clarification was of great value to Interfutures in that project's initial phase, when the various aspects of interdependence between developed and developing countries and their associated conflicts had to be reviewed.

Specific Difficulties in the Prospective Analysis of International Relations

The specific difficulties inherent in prospective analysis of international relationships stem from the fact that the system of international relations is one of the most global imaginable. The more global the system under investigation is, the less one can bank on the invariance of the structural relationships between the component subsystems and use exogenous assumptions concerning those relationships.

It seems that three levels of analysis are necessary in order to resolve the difficulty of gauging the dynamics of structural change in a global system.

The first involves *taking account of the interdependences* within the global system, whether these be interrelationships between the strategies of the various actors (governments, firms, multinationals, etc.), the links between the prospects for the different sectors considered (agriculture, energy, branches of industry) or the relationships between major macroeconomic functions (production, investment, trade, balance of payments). It is at this level that modelling, seen as an extra tool, can be particularly useful. Nevertheless, while the consistency of the global system imposes constraints on its various components, these components have their own, sometimes conflicting, dynamics which may influence the evolution of the whole and modify its structure. Prospective analysis of international relationships therefore poses in particularly acute fashion—because of the globality of the system under study—one of the essential problems in any prospective exercise, namely the dialectic between the evolution of the whole and that of its constituent parts, as a means of identifying possible structural changes.

The second level is that of *the strategies of the different actors* which may change the international balance of power and therefore the actual functioning of the system of international relations. These actors (nation-states, multinational companies, national socioeconomic groups) have projects and designs for the future which differ and often conflict. These conflicting projects take concrete form in medium- and long-term strategies. These strategies in their turn are powered by the specific forces (political or economic) which the actors wield in order to bring their pressure to bear on the international environment. Changes in the balance of power compromise the stability of the mechanisms which serve to regulate the system of international relations; new actors may emerge, others may see their roles dwindle. The actual structure of the overall system may thus sustain shocks which cause it to change.

Such an analysis based on the attempt to reveal tensions among actors, on the elucidation of conflicting projects and strategies, on shifting of balances of power, may seem biased towards some latent desire for structural changes to come about at all cost. This is not so. In fact, an analysis carried on in these terms may very well lead to the conclusion that structural changes are highly improbable and that the prevailing balance is maintained or even reinforced.

The third level of analysis is that of *prospective analysis* which may be termed *partial*. It may involve sectoral prospective analysis (population, raw materials, energy, agriculture, individual industries), or analysis of more horizontal questions (technology, monetary system, international trade). Clearly this type of analysis is quite fundamental inasmuch as it makes it possible:

(1) to pose in concrete terms specific problems which may give rise to policy formulation;

(2) to identify structural changes at global level which may be caused by trends in specific areas;

(3) to evaluate in concrete terms the forces commanded by the main actors involved, the constraints to which they are subjected and the strategies open to them.

It cannot be emphasized enough that these three levels of analysis are complementary. Each in its own way, world prospective studies, and in particular those mentioned above, give priority to one or preferably two of them, and this explains some of their most obvious drawbacks. It is because of its special efforts to integrate the three complementary levels of analysis that the Interfutures project is of interest from the methodological point of view.

INTERFUTURES' PROSPECTIVE APPROACH

We shall examine successively: the research options, the general architecture, the integration of quantitative and qualitative aspects and the role of the scenarios, the establishment of the analytical and historical base, the choice of the dimensions for the scenarios, the definition of the scenarios, the realization of the additional studies and partial prospective analyses, the elaboration of the scenarios, and the final presentation of results.

The Research Options

The foregoing remarks are enlightening as regards Interfutures' choice of research options. These options led to the development of an original methodology; as we have tried to show, they turned not only on the project's terms of reference but also on the way the team saw the specificity of prospective analysis and evaluated the knowledge acquired from numerous world future-oriented studies.

Obviously, despite taking the precaution of canvassing a broad range of views, the members of Interfutures' international team, with scarcely 2 years at their disposal (given the time necessary to launch the project) could hardly fail to be partially conditioned by their own disciplines, the institutions to which they belonged, cultural attitudes in their countries of origin and the atmosphere of the intergovernmental organization sponsoring the project.

Interfutures' research options can be described in a few sentences:

(a) An effort in prospective analysis. The outcome of this decision in favour of prospective analysis (as opposed to a forecasting exercise) was the construction of global or partial scenarios. These scenarios offer pictures of the future at a certain time-horizon and also describe, to some extent, the paths that led there. Over and above the analysis of possible changes, their interest lies in the light they throw on the possible policies of certain actors, particularly governments.

(b) An attempt to perceive both problems of interdependence and specific problems in a historical perspective such as will explain the actors' strategies. This option arises from the need, mentioned earlier, to take account of three complementary levels in analysis of the structural changes which may affect the system of international relations:

(1) It is essential to identify *interdependences*, since certain constraints are exerted at systemic level. However, such an approach, which generally uses world models to provide a consistent framework, has two major disadvantages. First, it proposes only an over-simplified picture of the world, one in which the various specialists do not recognize their own problems and which at the same time is too general to provide a base for decision-making by governments. Second, the modelling of interdependences, while it may identify evolutionary constraints, does not allow for the generation of structural changes.

(2) It is also essential to analyse long-term prospects in specific areas, whether sectoral or horizontal, since this makes it possible to integrate a considerable amount of information, to pose policy problems in concrete form, and to suggest possible structural changes which may originate in partial areas of the system studied and come to affect it as a whole.

(3) Lastly, some historical perspective seems to be indispensable, so as to give an idea of the developments which may be possible in the system under study as a result of the explicit or implicit strategies of the actors involved and the changes in the balance of power between them. Here, historical perspective offers a coherent qualitative framework.

(c) An approach which is primarily economic but accommodates political and social elements also. The future evolution of the world is a subject that can, and must, give rise to numerous interpretations and exploratory approaches. Political or military, cultural or social, economic or technological approaches are all legitimate, but they only form part of the whole.

In view of Interfutures' position in OECD and its terms of reference, the project's approach is predominantly economic. However, the more distant the time horizon, the more the analysis of the system of international relations necessitates a constant effort to integrate the economic, political and social phenomena which structure the system's development, without forgetting the factors linked with physical constraints.

(d) Research aiming at an understanding of the problems with which OECD governments will be confronted. According to its terms of reference, Interfutures has sought to identify the possible courses of action open to a specific group of actors, i.e. the governments of OECD countries.

In this framework, Interfutures has sought to build up a consistent and global picture, both qualitative and quantitative, of the various possible trends in the world economy and to reveal the long-term policy problems which require action from the OECD countries in the near future. This picture rests on the methodology which we will now set out.

The General Architecture

Figure 1 summarizes the approach. By describing first the project's general architecture, and then each stage of its progress, we shall show how this approach takes in the different levels of analysis previously mentioned.

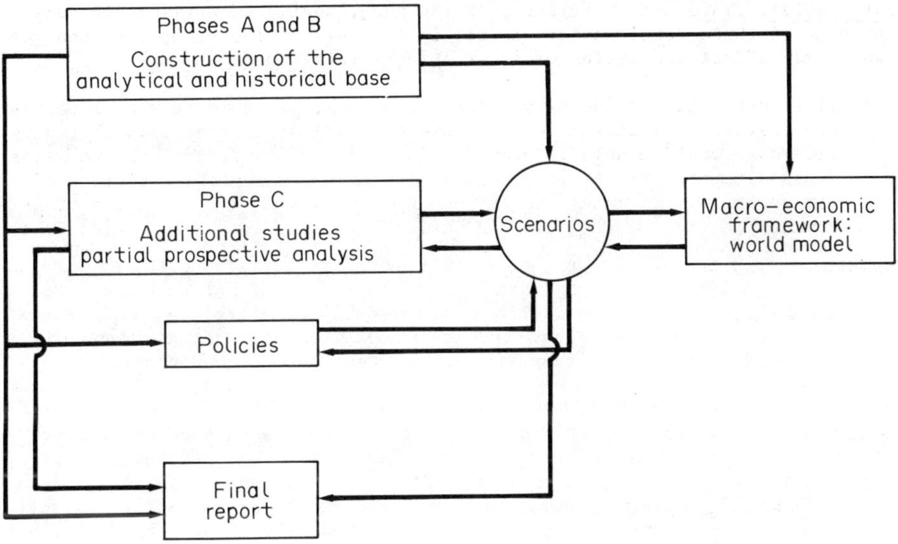

Figure 1. The general architecture of Interfutures.

The first stage of Interfutures (phases A and B) was placed in historical perspective and conducted simultaneously in terms of concept and of fact. It enabled the team to define *the general framework* of research by identifying the main subsystems (physical, geopolitical, economic) whose interacting dynamics over the long term were to be studied. It also made it possible to tackle analysis of *the specific problems* which are or will become issues or constraints as international relationships evolve.

This first stage was expected to yield a number of results: the identification of questions for additional study or partial prospective analysis; the demonstration of problems of interdependence which would then have to be expressed in terms of relations (formalized or not); the spotlighting of policy issues; and finally, a first definition of the dimensions or types of assumption which would govern the scenarios.

In the second stage, construction of the scenarios and partial prospective analysis were carried out in parallel. This essential phase made it clear that the scenarios played an integrative role: through the use of a world macroeconomic model they took account of interdependences; and they gradually incorporated in their sequential development the results of the additional studies and partial analyses.

The next step was to spotlight policies in so far as these could be directly deduced from the results of the additional studies and partial analyses and from the iterative process of scenario-building. The latter process made it possible, in particular, to take a second look, area by area, at the conclusions of phases A and B, and to reassess the plausibility of results.

The last stage was to draw up the final report presenting the main conclusions of the research.

The Integration of Quantitative and Qualitative Aspects and the Role of the Scenarios

Before describing the various stages of this process, it is worth dwelling for a moment on the essential role played by the scenarios in its organization.

The argument developed earlier suggests two conclusions as regards methodology.

First, quantitative exercises—and foremost among them, global models—should in no way be allowed to exclude or take precedence over other analytical methods. They should be used, not as a substitute to make up for the inadequacy of political economy when applied to international relations, but simply as what they are: tools to be employed in formalization, having their basis in explicit economic logic and in a consistent accounting framework which makes it possible to identify and describe interdependences.

Secondly, the temptation to make projections (induced by the use of quantitative tools) must be firmly resisted. Projection must not be considered as the principal aim of prospective analysis—far from it, and probably least of all in the study of international relations. The aim should be not so much to make quantified conjectures, expressed in terms of probabilities, about the possible long-term results of the interaction of various strategies, but rather *to use* such conjectures to identify the complementary or conflicting features of these strategies and the actual forces on which they can rely over a long period.

The scenarios, as they have been designed by Interfutures, have an integrative role—even if it is only a limited one—between the quantitative and qualitative approaches as well as between global and partial prospective analyses.

Firstly, on the basis of hypotheses as explicit as possible concerning the strategies, objectives and projects of the various actors—at national and international levels—there is an *a priori* identification of the types of international environments—i.e. systems of political and economic relationships—which may result from the confrontation or the complementarity of these strategies and projects. The various scenarios are designed to fit in with these types of international environment.

In a second stage the quantitative exercises and the models on which they may rely are conceived as the quantitative translation of the above-mentioned hypotheses. These exercises yield the main orders of magnitude associated with each of these scenarios.

Yet the modelling exercises can only translate on a formal level the dynamic processes leading to the potential realization of plausible end-state scenarios. This is why they must simultaneously be complemented by more detailed analyses, both quantitative and qualitative, concerning sectoral or horizontal aspects. These analyses, apart from their specific interest in explaining particular constraints or shedding light on specific policy issues, aim essentially at assessing the extent to which the strategies of the actors can effectively rely on *concrete* political, economic and technological forces.

Moreover, such analyses serve other purposes. On the one hand, being exogenous to the global models, they may help to identify inconsistencies or constraints which would remain hidden by the accounting equilibrium of the models. On the other they allow the assessment *a posteriori* of the plausibility of the starting hypotheses which characterize each of the international environments which are *a priori* defined.

As *analytical tools* the scenarios thus have an integrative role among various approaches, but beyond this instrumental role their purpose is to describe *the dynamic paths* as well as *the outcomes at the year 2000 horizon* which may result from the confrontation of the actors' strategies and some uncertain consequences of this confrontation. Therefore they do not constitute forecasts but rather attempts to shed light on possible or plausible futures. However, some of their features which remain stable in different scenarios may naturally appear as forecasts.

The significance of these long-term scenarios differs from that of most predominantly macroeconomic short- and medium-term scenarios. The main difference lies in the principle of consistency, which in these two types of scenario is not and cannot be applied in the same way.

In short- and medium-term scenarios there must be consistency both in the assumptions and in the paths followed. For the period under study, this is provided by the invariance of the model's structure or the conceptual frame of reference, and the end-state obtained cannot, in itself, call into question the starting assumptions.

In long-term scenarios, on the other hand, the assumption of structural invariance cannot and must not be taken as axiomatic. As the time horizon recedes, the accounting character of consistency is accentuated, to the detriment of its structural character. The integration of elements from quantitative and qualitative partial studies in working out the paths to be followed by the scenario leads to *the discovery of tensions*. Examination of these tensions, which are an obstacle to the realization of the scenario, may reveal either that some of the actors involved are going to have to change their policies, or that disruptions are possible—in other words, discontinuities such that, once they appear, developments may break right away from what was assumed as a starting point for the scenario.

While the scenarios are central to Interfutures' prospective approach, they are not its only end-product: the sectoral studies also reveal key important aspects of the future. Now that the part played by the scenarios has been established, however, we can examine in greater detail the stages already outlined.

Construction of the Analytical and Historical Base (Phases A and B)

As we have seen, this first stage aimed at two things: the constitution, through the study of specific subjects, of a factual analytical base; and the establishment of a general framework in order to conceptualize the international relations system; to identify the main actors; to investigate peripheral problems which, while they are outside the project's terms of reference, have to be taken into account

This framework was substantialised by five papers:[1]

(1) a first assessment of world problems;
(2) a comparative evaluation of world models;
(3) an analysis of the evolution of international relations during the past quarter of a century;
(4) a study entitled "East, West, and South: the impact of Eastern Europe";
(5) an investigation of the impact of armaments on longer-term development.

These studies, particularly the one on international relations, were of special importance in the establishment of the consistent framework without which no scenario can be built.

The investigation of specific questions made it possible to study the nature of the issues involved from a sectoral point of view. In examining the sectoral prospects the research team generally went only so far as to identify the points at which there was interdependence, without analysing the interdependent relationships in detail. The main aim in this phase was to identify the constraints to which the various actors were subject and the forces available to those actors in carrying out their strategies in each area. The research also sought to identify, in each area, some of the strong trends liable, over the long term, to extend to the whole system and thus apply to every scenario.

Studies were carried out on population, energy, nutrition and agriculture, raw materials, the international division of industrial labour, the future of technology in advanced industrial societies, the problems of technology transfer between advanced and developing societies, physical environment, international trade, long-term aspects of the monetary system, and finance issues.

In addition, five other documents drafted during this first stage began the integration of the studies on specific subjects into the general framework and at the same time prepared the move from the research phase to the policy-definition phase. Four of these dealt, respectively, with the internal dynamics of advanced industrial societies; patterns of evolution of developing countries; institutional problems; and dimensions of alternative long-term scenarios. The fifth gave an item-by-item summary of issues for the advanced industrial societies.

Once the general framework had been set up and prospects examined sector by sector (or area by area), then the actors' projects and the constraints they might face could be identified and the scenarios assembled as contexts for the analysis of possible conflicts or tensions between the strategies associated with these projects, and as a means of shedding light on the policies which might help resolve these conflicts and tensions.

It is significant that phases A and B were concluded by the definition of the dimensions of the scenarios.

The Choice of the Dimensions of the Scenarios

The interest of the scenario method lies in the fact that it allows the conceptualization of structural changes which may affect the evolution of the system under study. *The dimensions of the scenarios are sets of hypotheses which account for the main possible structural changes.* Naturally, the process calls for simplification consistent with the representation chosen for the system studied.

It is not surprising that the three phase A and B studies most inspired by a historical perspective—those on the evolution of international relations, the internal dynamics of advanced industrial societies and the patterns of evolution of developing countries—were the ones which led to the definition of the scenarios' dimensions.

The first of these analyses sheds light on the *changing balance of power, on the one hand among developed countries; on the other between developed and developing countries,* without forgetting the role of Eastern countries.

The second one tracks down *the possible changes in the pattern of growth in advanced countries.*

The third one considers *the possible strategies open to the new actors* formed by developing countries in order to change an international order in the establishment of which they have not participated.

In fact—and this is rather logical—these analyses reflect the three main aspects of the crisis which characterises the present 10-year transition period, the structural crisis within the developed countries, the crisis of their mutual relations, the crises in the political and economic relations that link them with the LDCs.

Thus, the dimensions which are used to define the scenarios are essentially conceived to reflect *the uncertainties concerning the possible outcomes of the three aspects of the present transition period.*

For each dimension, Interfutures defined contrasted hypotheses which seem the most significant, either because they correspond to specific strategies by some actors or because they represent possible outcomes of existing tensions. Each end-state scenario is characterized by a combination of these main hypotheses. This approach is interesting because the choice of dimensions and their associated hypotheses means that the "space" of possible futures can be demarcated.

Four dimensions seem to be the inevitable minimum to map this "space" of possible futures. Each covers contrasting assumptions which are simply listed here without any attempt at justifying them.[2]

The first is the nature of the relations that may establish themselves between developed countries (North-North relations). Two intermediate assumptions represent two kinds of possible outcome to the problems posed by the rebalancing of economic relations within the developed world.

(1) Within the group of advanced industrial societies, the trend of relations favours a transition to *collegial management* of the developed world's interests by the largest countries or by the United States, the European Economic Community and Japan. This collegial management would ensure a relatively high degree of openness and some measure of economic

stability.

(2) On the other hand, *conflicts* between advanced industrial societies in specific areas will lead to *the partial abandonment of free trade*, with the United States, Japan and the EEC as competing poles.

The second dimension concerns essentially the nature of relations between developed and developing countries (North-South relations). Three assumptions cover the range of possibilities:

(1) Progressive integration of the Third World (albeit with internal disparities) in the world markets as a result of *substantial increases* in the flows of international trade and capital.

(2) *More pronounced dissension between North and South*, with the Third World countries attempting a "de-linking" strategy.

(3) *A regional break-up of the South* caused by the formation of groups around emerging centres of power, in alignment with developed country poles at the centre of zones of influence.

The third dimension relates to the *internal dynamics peculiar to the developed societies*. The aim is to provide a description of changes in those societies which differ in depth and which result on the economic plane in social products differing in *composition* or *volume*. This dimension is expressed by the two-fold assumption of productivity growth and changes in values.

Crossing the alternatives concerning growth rate and changes in values produces *a priori* three assumptions:

(1) Priority given to strong economic growth in line with past trends, with no major change in values.

(2) Social fragmentation with changes in values not unanimously accepted and an exacerbation of the conflicts between social groups. Slow growth over a long period.

(3) Rapid changes in values, particularly with regard to work and the composition of the goods and services demanded. Consensus on slow growth and a different content of national income.

The fourth dimension naturally relates to the internal dynamics specific to the different groups of developing societies. Here again, it is by combining *a rate of growth of national income* and a *development strategy* assumption that an attempt can be made to identify various possible lines of evolution for groups of Third World countries.

On closer investigation it could be seen that development strategies in these groups of countries very much depend on the nature of political and economic relations with the developed countries. In other words there is a strong correlation between the second and fourth dimensions of the scenarios. Because of this, when the scenarios were constructed the fourth dimension was made endogenous and in the case of the various groups of Third World countries an attempt was made to deduce the potential growth rates and development strategies most compatible with the assumption underlying the second dimension. As will be shown later, this was made possible by partial prospective studies on a number of developing countries.

Before specifying the choice of combinations of assumptions that define the scenarios studied by Interfutures, it is necessary to consider two fundamental points which lie outside the foregoing dimensions. The first is the evolution of *relative* productivities in the developed world. The second is the impact of *Eastern Europe*.

Obviously the evolution of relative productivities in the various parts of the world, and particularly among the developed areas, has a direct effect on relative competitiveness and therefore on trends in the balance of power. Two alternative possibilities were envisaged. The first supposes convergence of growth rates and levels of productivity in the main developed areas between now and the end of the century. The second supposes divergence and is based on trends in the past where, in the long run, changes in the order of the main developed countries' productivity levels have occurred.

These two pictures of productivity evolution are linked mainly to the third dimension of the scenarios. However, the assumptions adopted in connection with the first and second dimensions are bound to have a differential influence on productivity, as the scenario-building process will show.

In defining the scenarios no direct account was taken of possible developments in Eastern Europe and their impact on the world system, largely because the project's terms of reference were limited. Exogenous assumptions are made concerning economic and political trends in the Eastern European countries. These assumptions are integrated quantitatively at the level of the world model which provides the consistent accounting framework for the scenarios; the qualitative consequences, on the other hand, particularly as regards the assessment of the scenarios' plausibility and stability, are examined *a posteriori*.

The Definition of the Scenarios

Each scenario is defined by an identifiable and reasoned combination of assumptions. Of all the combinations possible, some appear more plausible than others, notably because of the interdependence of the dimensions. But *a priori* plausibility is not the only criterion. Some scenarios may be pertinent because they reveal the incompatibility of strategies pursued by protagonists in a situation of conflict, or because they develop trends which as yet are embryonic.

In trying to reduce the combination of assumptions to a minimum, while at the same time keeping the range of futures envisaged open, Interfutures investigated in greater or lesser detail the scenarios summarized in Table 2.

Scenario A Collegial management of interests and conflicts in the developed countries; increased free trade; increasing Third World participation in world economic exchanges, but varying as between developing countries; sustained economic growth in the developed countries but no rapid change in values. Relative productivities in OECD countries are assumed to converge.

Scenarios B1, B2, B3 Identical assumptions as to the nature of relations between developed countries, between developing countries and between the two groups. On the other hand, the developed countries will experience only moderate growth with differences according to the three alternative scenarios. In B1, value changes are rapid and there will be a consensus on the slowdown in growth, since it will be accompanied by a change in the content of "social output" (in the broadest sense of the term). In the other two

TABLE 2 Definition of the Scenarios

Relations between developed countries	Collegial management		Partial abandonment of free trade between the poles
Internal dynamics of the developed societies	Consensus in favour of high growth	Rapid value changes and moderate growth	Conflicts between social groups and moderate growth
Trend in productivities	Convergence		Divergence
North-South relations and relations between LDCs			
Large growth of North-South economic exchanges	A B1	B2	B3
Accentuation of divisions between North and South			C
Partial fragmentation of the South following regional alignments with the developed country poles			D

alternatives, however, there is no significant, unanimously accepted change in values and the slowdown in growth is due more to structural adjustment difficulties at national and international level than to conscious resolve as in B1. Whereas the B2 alternative supposes convergence of relative productivities, B3 assumes divergence due to the social and institutional disparities between the developed countries.

Scenario C This is a scenario of North-South confrontation. It supposes the implementation of "de-linking" strategies by a majority of developing countries, collegial management by the countries of the North with increased liberalization of their trade, slower growth without any change in values in those countries, and no convergence of productivity because the main OECD zones would be variously affected by the North-South break.

Scenario D Break-up of the developed-country group and mounting protectionism with the emergence of zones of influence centred around three poles: the United States, the European Economic Community and Japan. These zones will include regional groups (on a continental scale) of developing countries;

trade and capital flows will develop preferentially within those zones. These assumptions are coupled with that of slower growth due in part to the destabilization of trade flows. Non-convergence of productivities is due here to the differing impact of the break-up process on the main OECD zones.

Despite the diversity of these scenarios, the picture of the possible futures they propose is obviously not complete, for several reasons.

First, the project deliberately excluded scenarios supposing major upheavals such as sudden large-scale reduction of energy supplies, radical changes in the sociopolitical systems of the main developed countries (of either East or West), military conflicts directly involving the great powers. *The exclusion of these crises in no way signifies that they are thought to be impossible; they are taken into account in the final report—but not through the medium of the scenarios.*

Next, because the scenarios envisaged depict extreme cases. They make it possible to cover a whole range of possible futures. But the real future—unless major crises do occur—will probably be a composite of elements of the different scenarios.

The Realization of the Additional Studies and Partial Prospective Analysis

The scenarios, as defined so far, give very global pictures of what the future might be. In this context, the additional studies and partial prospective analyses have three aims:

(1) to identify the structural changes which may be caused by trends in specific areas and, independently of the scenarios, to throw light on policy choices;

(2) to assess in concrete terms the internal and external constraints to which the actors involved in the various scenarios are subject, and the forces at the disposal of these actors for the implementation of the strategies open to them;

(3) to make these strategies explicit in terms of policy choices.

Interfutures undertook four kinds of additional study or partial prospective analysis, and to these must be added some of the specific analyses carried out in phase B.

(a) The additional studies.
These deal with *changing values* in the advanced industrial societies, and the *structural problems* which condition growth prospects.

Changes in values do not appear out of the blue. They are bound up with economic and social trends and furthered by social groups whose importance, motivations and characteristics are induced by structural changes; they cannot be understood except by reference to these groups. Similarly, and symmetrically, values are at the root of the behaviour patterns which facilitate or hinder structural change.

The main themes investigated, therefore, were the conditions in which new values emerge, their consequences for trends in social demands and their effect on the structure of the socioeconomic system. The alternative scenarios include assumptions on changing values at a very general level. The purpose of the additional study was to determine how far the correlation between changing values and changing structures was compatible with the logic inherent in each scenario, and to see whether this correlation strengthened or, on the contrary, reduced the scenario's plausibility.

The second additional study combined a detailed investigation of *the prospects for macroeconomic growth* with an analysis of the *pressures* to which the advanced industrial societies will be subjected in the future and of the *rigidities* which may make it hard for them to adapt themselves. Numerous questions had to be tackled, among them:

(1) macroeconomic aspects: the technological limits to growth, macroeconomic and social constraints on strong growth, prospects for moderate growth, population;

(2) pressures on structures: population, changes in final demand, the impact of energy prices and environmental protection costs, changes in the competitive positions of the developed economies both among themselves and with regard to the Third World;

(3) structural rigidities: population, the organization of the labour market, market economy-State intervention dualism, protectionist tendencies.

Study of these subjects yielded substantial information for the scenarios, and made it possible to discover the nature of the elements which might hinder or, on the contrary, facilitate the realization of a particular scenario. The plausibility of each scenario could thus be more accurately assessed, but more important it was possible to construct "composite" scenarios, these being plausible combinations of original scenario elements. For instance, the high-growth scenario became much more questionable following the introduction of macroeconomic constraints as to profitability of capital, balances of payments and inflation and changes in competitive positions endangering freer trade.

When the two additional studies were considered as a whole, they made clear the dangers of three extreme strategies which the developed countries might be tempted to follow: growth at any price, defence of existing structures, or "new growth"; they also made it possible to identify choices which somehow combined policies based on these three extreme strategies while restricting their adverse effects or irreversible consequences between now and the end of the century.

(b) The partial prospective analyses.
These covered two areas in particular: first, prospects in certain *industrial sectors* and second, *prospects for the evolution of developing countries or groups of developing countries*.

The importance for international relations of *developments in the international division of industrial labour* is now universally recognized. The Interfutures phase B study had shown that a more detailed and more concrete sector-by-sector analysis was needed, for two essential reasons:

First, treatment of industrial questions in the future-oriented world studies available to Interfutures was, on the whole, unsatisfactory. Either the degree of disaggregation was insufficient or the modelling approach entailed an over-mechanical treatment which masked the real levels of decision and the voluntaristic attitudes on industrialization which might be adopted by the actors involved.

In the second place, some of the existing or potential conflicts between developed and developing countries or among the developed countries themselves concern production of and trade in manufactured goods. Only sectoral analysis can reveal the nature of these conflicts and the forces at the disposal of the different actors involved.

Interfutures carried out eight sectoral analyses, on the following industries: non-ferrous metals (copper and aluminium), iron and steel, chemicals, capital goods (except transport equipment), electronics, shipbuilding, automobiles, and textiles. These sectoral studies were not, in general, carried out in the traditional manner, i.e. linking world demand prospects (disaggregated as by region) with supply prospects by calculating comparative advantages at production level, taking into account exogenous factors such as possible technological developments, the size of domestic markets or the degree of protectionism.

While these factors were taken into account, the approach used relied on the identification of actors' strategies *at the level of the sectors analysed, in terms of production processes*. In each case, an attempt was made to determine the advantages available to the protagonists in controlling trends in the international division of labour (upstream and downstream integration, technology, segmentation of production processes, investment abroad, and so on). In parallel, the research sought to identify the constraints on each country's productive system which might be created by the strategies of outside actors (States or multinational firms) and which might elicit various types of response: abandonment of some parts of the production process, positive action to improve technology and organization and reduce costs, protectionist measures, etc. An important advantage of this method is that it may lead to a reassessment of the plausibility of the scenarios on whose assumptions the estimates of the studies are partly based.

The same is true of *the prospective studies on developing countries or groups of developing countries*. Such an exercise was obviously necessary, since Interfutures' assignment was to study the future development of the OECD countries in harmony with that of developing countries; moreover, assumptions consistent with the possible strategies of various components were explicitly built into the alternative scenarios. However, the main reason for the necessity of analyses by country or group of countries was the diversity of the circumstances and development potential of the Third World countries and of their commercial, financial, technological and political relationships with the North. Such diversity could not be accommodated in too global a view of North-South relations.

Obviously, only a limited number of countries could be studied individually, and these were selected either because they are of major importance in terms of population or production volume, or because they offer typical examples of various development strategies. The countries chosen were the following: Algeria, Egypt, Iran; Tanzania, Nigeria; India, Pakistan, Bangladesh; Indonesia, the Republic of Korea, the People's Republic of China; Mexico, Venezuela, Brazil. In addition, further studies assessed prospects and strategies for integration into the world economy for the following regional groups: North Africa and the Middle East; Sub-Saharan Africa; South Asia; South-East Asia; and Latin America.[3]

The aim of these prospective studies by country or by region was twofold:

(1) to throw light on the long-term development prospects of a series of typical countries, by reference to each country's past performances, the

principal constraints confronting it and the main alternative strategies open to it;

(2) to identify the policy choices open to the OECD member governments in view of the prospects and development strategies of the various Third World countries.

The studies on the prospects for developing countries and groups of countries were integrated into the analysis in two ways:

(1) directly, where they showed the specific issues to which some types of strategies may lead (as regards international trade, technology transfers, capital flows);

(2) indirectly, where they made it possible to assess the plausibility of the scenarios.

The Elaboration of the Scenarios

We can now go into more detail concerning the integration process outlined in the section on the project's general architecture. This will be done by using as an example one of the actual Interfutures scenarios and showing, step by step, the operations involved in building it.

Before embarking upon this description, however, it may be advisable to pause for a moment to consider the use which was made of various models—particularly SARUM—in building the scenarios.

The use of models. SARUM was developed by the Systems Analysis Research Unit (SARU) of the United Kingdom Department of the Environment. It functions essentially through changes in demand and prices, and explicitly takes into account scarcities of resources in the agriculture and energy sectors.

Some changes in the model were made by SARU in collaboration with Interfutures. They consist, in particular, of an increase in the number of regions and an extension of sectoral disaggregation (12 regions, 11 sections) which make it possible, in analysing industry, to separate capital equipment from other manufactured products.

In particular, technological progress was partially incorporated by making the effect of vintage capital endogenous, like that of capital flows on indebtedness and balances of payments. Fundamentally, however, the logic of the model, at the level of each economy, remains neo-classical. Production is assumed to be carried out by sectors of firms seeking to minimize costs, and decisions on consumption and savings are taken in such a way as to maximize a utility whose structures differ from one region to another. Production processes are represented by Cobb-Douglas functions, with an exogenous component for technological progress.

One of the main reasons for choosing the model was its treatment of international trade, which offered a distinct advantage by making it possible to simulate the effects on trade of tariff and non-tariff barriers. These are introduced into the model in the form of "bilateral trade biases" matrices, which are applied in the determination of trade flows. The bias takes the form of notional increases in the prices of traded goods. They are fixed initially so that the trade observed in about 1970 is reconciled with that which would have taken place with the observed price differentials in a free market. By increasing or reducing these trade biases it is possible to simulate increases or decreases in trade barriers.

Although SARUM is the central model which ensured the accounting consistency of the scenarios and made it possible to estimate the macroeconomic or sectoral orders of magnitude associated with them, other models were used as well. We should mention, especially, an international trade model which, using the same macroeconomic assumptions for the scenarios as those which served as input to SARUM, measured the effects on trade flows of changes in price and income elasticity, and also the relations between balance of payments and real exchange rates.

Steps in scenario-building. We shall describe here the successive steps in building Scenario A.

The qualitative assumptions corresponding to the scenario were partly quantified so that they could be incorporated as input into SARUM. For Scenario A, the *growth path of the OECD countries* is the result of a calculation which assumes simultaneously:

(1) The making good of losses in production and/or productivity due to the recession. This catch-up takes place by 1990, the date on which unemployment is assumed to be reduced to its frictional or full employment component, which will vary from one country to another.

(2) Convergence of the productivity trends of the various countries on the long-term trend in the United States, determined exogenously.

(3) Changes in population and participation rates determined exogenously by extrapolation of recent trends.

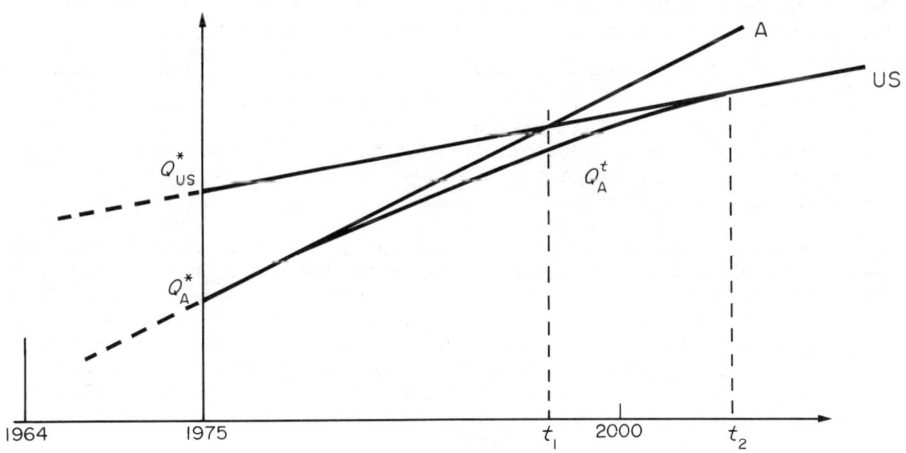

Figure 2. The growth path of OECD countries

If: Q^*_{US} is the potential productivity of the United States in 1975, defined in the following way: productivity obtained by prolonging the trend between the last 2 years of the 1964-73 decade of minimum unemployment,

Q^*_A is the potential productivity of country A, in 1975, $S_{US}(t)$ is the long-term productivity trend of the United States, determined exogenously, $S_A(t)$ is the trend of *potential* productivity of country A extrapolated from the figures observed for the period 1964-73, U^*_A is the frictional component of unemployment in country A. LF^t_A is the labour force of country A in year t, Q^t_A is the estimated productivity of country A in year t,

One assumes: $Q^t_A = Q(Q^*_{US}, Q^*_A, S_A(t), S_{US}(t), t)$, Q being the function which ensures convergence of productivities in year t_2 so that $(t_2 - 1975) = 1.5\ (t_1 - 1975)$ (see Figure 2).

The productivity trend for the period having been determined, and taking into account assumptions on population, participation rate and unemployment, the gross domestic product for country A in 1990 and in 2000 can thus be deduced:

$$\mathrm{GDP}^{1990}_A = Q^{1990}_A \times LF^{1990}_A\ (1 - U^*_A)$$

$$\mathrm{GDP}^{2000}_A = Q^{2000}_A \times LF^{2000}_A\ (1 - U^*_A)$$

GDP in 1975 is known; the trend of GDP between 1975 and 2000 will be obtained by smoothing the trend between these three points.

The calculation process,[4] which is the same for all countries, may naturally result in more or less realistic figures depending on the country, but so far as the main economic centres are concerned, this does not affect the significance of the scenario.

The growth figures for *the developing countries* over the period 1975-2000 are based partly on the analyses by country or by region. Taking into account the assumption of integration into the world market, growth elasticities for the developing regions are estimated in relation to growth in the developed world.[5]

Lastly, for the Eastern European countries and China, the rates of growth are determined exogenously; they are based, for Eastern Europe, on the forecasts of the United Nations Economic Commission for Europe and for China, on the Interfutures study of that country's development prospects.

There are three other sets of supplementary assumptions:

(1) This concerns the elasticity of demand for energy in relation to growth. It is assumed, in the light of the phase B study on energy, that this elasticity will be reduced in both developed and developing countries.

(2) This specifies the degree and rate of trade liberalization in a form which can be taken into account by the world model used.

(3) This concerns the mechanisms for external financing of the payments deficits of the developing countries. Assumptions have been introduced regarding increased aid and its allocation, and also direct investment and loans.

With this group of assumptions introduced into the model used, it is possible to define a path and an end-state at macroeconomic level and at a sectoral level.

Once this path and end-state, consistent from the accounting point of view, have been determined, the next step is to introduce the results of the additional studies and partial prospective analyses (by sector and by developing region). As it takes on more substance, the scenario at the same time undergoes what might be called "criticism from within". It keeps its accounting consistency, but gradually loses its structural coherence as the qualitative or quantitative elements from the additional studies and partial prospective analyses challenge the initial assumptions at the level of the paths followed and of the end-state. The studies introduced reveal the tensions which might prevent the actual realization of the scenario, and suggest policy choices which might help to resolve those tensions.

During this process the possible crises and the policies to counter them come under review: the consistency of the macroeconomic image for the developed countries, given the elements which become apparent upon analysing the interplay between values, growth and structure; the energy problem; world agricultural production and trade; the international division of industrial labour; international trade; capital flows.

A synthesis of all the questions posed then shows the conditions necessary for the realization of the scenario and also reveals its instabilities. Simply to illustrate this methodology, we give below, without comment, the elements of the synthesis made for Scenario A.

The realization of this scenario assumes that many obstacles are overcome. For the developed countries, the *internal* and *international* aspects of those obstacles are closely intermingled: sustained and regular recovery in investment, acceptance of continuous structural adjustment, concerted international action in the management of overall demand, trade liberalization, massive increase in financial flows with regard to the Third World, and responsibility in energy matters.

Underlying these economic conditions, however, the scenario poses *political problems* about which it is impossible to avoid asking questions: the changing role of the United States, the growth of Japan, the cohesion of the EEC, and participation by Third World countries (OPEC countries, low-income countries. These problems are discussed in the final report.

The Final Presentation of Results

Scenario-building is at the heart of the Interfutures approach. The scenario method has the advantage that it can be used both for analysis and for shedding light on policies. It is a conceptual "detour" through plausible futures fostered by the strategies of the actors involved as they unfold through time. However, as the diagram on general architecture shows, the integrative effect of the scenarios is only *partial*. Constraints or persistent trends which will induce policy choices can also be identified by studies upstream from the scenario-building (phases A and B of the project), or by additional studies or partial prospective analyses which feed new elements into the scenarios.

It is for this reason that the plan chosen for the presentation of the project's final results gives pride of place to the major problems which are bound to lead to policy formulation, with the sections on scenarios taking second place. Thus the Interfutures final report is in five parts:

Part I, without adopting any position as to the legitimacy of economic growth, tackles the essential preliminary question of the physical limits to this growth.

Part II, in the light of the quantitative results obtained through the scenarios, this gives some indication of the range of the main orders of magnitude which may affect world economic development between now and the end of the century.

Part III deals with the long-term prospects for the advanced industrial societies and their capacity to adjust to structural change in economic, social, institutional and political terms.

Part IV concerns prospects for the Third World and the evolution of relations between developed and developing countries.

Part V, which includes a long section on the scenarios, is devoted to the problems inherent in the medium- and long-term management of world interdependence.

III. CONCLUSION

In conclusion, we shall try, first, to make clear the limitations of the Interfutures prospective study, whether these are inherent in the method used or due to inadequacies in putting it into practice. Then we shall look at the lessons which world prospective studies can draw from the project.

The Limitations of the Study

It would, of course, be quite wrong to claim that the methodology worked out by Interfutures answers satisfactorily all the epistemological questions detailed in the introduction. Certainly, from the moment the ambitions of prospective analysis have been set out, there is a sense of impoverishment which builds up throughout the critical review of the kinds of investigations needed in order to serve these ambitions in a study of international relations, and culminates in the description of the methods used by Interfutures.

But rather than trying to show up the weaknesses of the method against the ambitions of prospective analysis, it is probably more useful to emphasize the insufficiencies of its application and, given its aims, some of its inherent limitations.

The *insufficiencies of its application* are largely due to the short time available. Whereas the project as a whole lasted something over 3 years, the time devoted to research was no more than 2 years, given the time necessary to get the team together at the beginning and to draft the final report at the end. In these circumstances:

(1) It was difficult to complete phase A and B studies when it became apparent, at the time of their integration into the analysis, that certain phenomena had been underestimated.

(2) It was not possible to review the partial prospective analyses after the figures for the scenarios had been worked out so as to examine in detail the consequences of the scenarios for the various sectors.

(3) The list of scenarios had to be decided fairly early on, and once this first series had been studied it proved to be impossible, for lack of time, to add others or to modify the underlying assumptions.

(4) Originally, the programme called for the use of several models; the framework provided by SARUM was to have been complemented and made more detailed with the help of more specialized models. For lack of time, the only one used, for Scenario A, was a model on trends in real exchange rates.

(5) Finally, the study on breaks or discontinuities, which had been planned from the start, could be undertaken only at the end of the research work and so was not as thorough as would have been desirable.

By and large, however, the methodology used proved suitable, and the additional studies considerably improved the structure of the project as a whole, without affecting the form of the final report.

Then again, the intergovernmental framework in which the research was carried out proved to be an undeniable advantage, but at the same time a disadvantage: the former because information was readily accessible, because there could be contacts with senior civil servants from many countries, because it was easy to obtain the collaboration of a large number of experts from both the developed and the developing countries; the latter because it undoubtedly slowed down the work (due to language problems, the fact that even working documents had to be subjected to a certain amount of censorship by the writers themselves, delays in feedback). While the team benefited from the wealth of variety offered by recruitment from many nations, its members had at first no common approach to prospective analysis. This approach could not be decided arbitrarily, but took form gradually as the research progressed. The Steering Committee which supervised the project and consisted of senior civil servants (one representing each participating country) was naturally somewhat concerned, at times, by the project's complex approach, which widely overstepped the traditional bounds of applied economics. Only when the preliminary version of the final report became available did all the participants fully grasp the significance of the method used.

The main *limitation of the method relative to its aims* concerns one aspect of the role of the scenarios. As a first step, starting scenarios are defined by means of a limited number of dimensions. The range of these scenarios is intended to delineate a "cone of possible futures"; but, in fact, only a very small number of dimensions can be given. There are three main reasons for this:

(1) The combination of assumptions associated with the dimensions, if their number is large, can very soon lead to an unmanageable quantity of scenarios.

(2) The assumptions associated with the dimensions have to be quantifiable so that they can be incorporated into the models which ensure accounting consistency.

(3) The starting scenarios must remain flexible enough in their descriptive characteristics for the data provided by the additional studies and partial prospective analyses to be integrated progressively.

The fact that the number of dimensions is limited thus entails, in practice, some closing of the cone of possible futures, so it is dangerous to adhere to the assumption that the future will contain elements of the extreme scenarios and—to sustain the metaphor—still fit inside this cone.

Another criticism is that the scenarios as they stand *do not permit more than a superficial treatment of possible "breaks"*. It is true that by definition, one might say, assumptions such as armed conflicts or radical changes in the sociopolitical systems of the OECD countries, which would cause irreversible breaks, were discounted at the outset. But the very fact that a world model is used does away with any chance of analysing certain possible disruptions (and this remark especially applies to SARUM, given its neo-classical characteristics). Over and above the use of models, another matter for criticism is the fact that as soon as a break is foreseeable, the logic of the scenarios tends to emphasize the regulating mechanisms. Conditions in which breaks might occur are indeed mentioned and even analysed, as in the case of energy, the exacerbation of conflicts between the developed countries or the "de-linking" of the Third World; but following the method's logic, the aim is to shed light on the strategies actors faced with these possible breaks would adopt, rather than to imagine the consequences such disruptions, if they happened, would bring in their wake.

In conclusion, let it be said that any prospective study must suffer from the limitations built into its methodology, and Interfutures, an ambitious project, had to pay the price of its own aims: to marry quantitative and qualitative analysis and to combine with certain method components of unimpeachable accuracy other components based essentially on intuition and sociopolitical judgments. What lessons can be learned for the future?

The Lessons of the Study

If there is one feature which is specific to the long-term evolution of social systems, it is the interdependence of political, economic, technological, social and cultural determinants. Any prospective study emphasizes, whether explicitly or not, a particular interpretation of future evolution. The Interfutures interpretation has been predominantly economic; however, a constant effort—sometimes successful, sometimes less so—has been made not to neglect other aspects either in themselves or in their economic connotations. It is safe to say that any future-oriented world study which tried to single out only one aspect and to endow it with a kind of autonomy enabling it to shape the pattern of social dynamics as a whole would be not only lopsided but illusory.

Like a jigsaw puzzle where the picture which will take shape is not known beforehand, world prospective analysis takes patience. If the many and varied investigations and partial studies which go to make up the overall methodology are truncated, then the pieces will not fit together. In this respect there is little doubt that Interfutures' work suffered because of the time limits allotted it. World prospective analysis requires sustained effort over a fairly long period if it is to be fruitful. Moreover, its methods and orientations are bound to be influenced by the institutions which undertake it. Here there are two eventualities, each equally dangerous for prospective analysis applied to international affairs: first, if the subject came to be monopolized by one or two permanent research teams; and second, if work were to be fragmented so as to consist of numerous short-run activities being carried out in scattered research centres having no contact with one another.

NOTES

1. Part of a larger, 21-chapter unpublished working document, "Midway through Interfutures", in which the results of phases A and B were presented.

2. These assumptions are discussed in more detail in Part II of the Interfutures final report.

3. All the additional studies and prospective analyses of industrial sectors were done either by Interfutures research team members or by experts from OECD countries. On the other hand, patterns of evolution of the developing countries, or groups of countries (with the sole exception of the study on the People's Republic of China) were analysed by independent experts from the countries concerned. This collaboration was found particularly useful since the research team members gained a different insight into the internal and external constraints on the Third World's development from that normally possible for OECD experts.

4. The method of calculation, and not only the value of the parameters, varies from one scenario to another.

5. The calculations take into account similar estimates made by the World Bank for a shorter period.

FAST: Forecasting and Assessment in the field of Science and Technology

P. C. Roberts

THE ORIGIN OF FAST

In 1974 a feasibility study was commissioned to investigate the case for a long-term futures group to conduct forecasting and assessment activities on behalf of the European Community. This study, termed "Europe Plus 30", was headed by Wayland Kennet[1] and drew on a group of distinguished European scholars. They recommended the creation of a substantial and independent institution, embracing a broad range of capabilities, so that it could offer to Europe's decision makers an integrated view of threats, opportunities, goals and policy options in the following fields: climate, population, agriculture, health, society, education, science and technology, industry, energy, materials, environment, transport, communications, economics and finance, defence, and politics and institutions.

To this ambitious recommendation, the response of the Commission of European Communities and Council of Ministers was cautious but positive. In July 1978 it was decided to establish a 5-year pilot programme in the field of science and technology, with a total staff (scientific and supporting) of 10. At the time of writing, the FAST (Forecasting and Assessment in the field of Science and Technology) team has been in operation for 4 years. The remainder of this report is concerned with the philosophy, methodology, and programme of the team.

FORECASTING

In order to appreciate the philosophy of the FAST team it is necessary to put their ideas on forecasting into a broad perspective covering the full range of viewpoints and techniques. At one end of this range is the "detached observer" seeking a small set of general principles through which the evolution of a system can be understood and its future state predicted. Examples of this style are seen in astronomy and weather forecasting. At the other end of the range is the "planner". He has all aspects of his project under control. His vision of the future is made to materialize—he wills it and it comes to pass. Examples are engineering projects such as bridge-building or space vehicle launching.

Intermediate in the range is the case of the "strategist" or "games player". He has some influence on the outcome of events but that influence is not decisive. There is uncertainty present in the constraints imposed by Nature; i.e. he is not sure what is possible and, in addition, there are other players in the game whose actions are not accurately predictable. He sees his task as that of studying the game and the players so as to achieve some prescribed set of goals through subsequent play.

For completeness, I define a further role—that of the disinterested analyst who differs from the strategist in that he is not confined to a set of prescribed goals, but embarks on a general exploration of potential outcomes.

The FAST philosophy is that of the strategist, though called by the team "prospective studies". The player in the game is the European Community, regarded as a single entity. Although there is a group of nations constituting the community, differences within this group are considered in the same way that a strategist serving a national government considers the separate interest groups within his country—groups such as ethnic minorities, professions or academia. The problem is seen as one of uniting the groups in a common purpose, in spite of the differences which exist among them.

The use of the term "game" conveys the idea of competition, together with the existence of winners and losers; but this is too narrow a definition of "game". We define a zero-sum game as one in which the total value of the payoffs is unchanged, i.e. whatever the winner wins must be the same as the loser loses. However, there is a class of non-zero-sum games in which the total value of all payoffs is affected by the chosen strategies of the players. In particular there are mixed modes of cooperation and competition which yield more to the player than if a purely competitive mode were selected.

The foregoing paragraphs define the style of the FAST team, distinguishing it from detached observer, planner and disinterested analyst. Three other characteristics are needed to identify FAST and these are:

(1) the areas of concern, i.e. the nature of the problems;
(2) the goals which guide the strategy;
(3) the techniques used in the formulation of the strategy.

AREAS OF CONCERN

For a group working within the EEC bureaucracy and seeking consensus for its objectives in order to justify its existence, the source of selection for the concern areas must be received wisdom. A compact summary is offered in "European Perspectives":

> In recent years, we have been confronted with a series of monetary, energy and economic crises. Each time it was necessary to wait until the crisis became a major one before action was taken. In some cases we still await a response. The acceleration of technological, social, political and economic change makes future disturbances increasingly probable and increasingly close to us. The only alternative, and this is precisely one of the essential missions of the FAST programme, is to list the promising opportunities and to identify the problems before they become critical. Only then can the appropriate R&D actions be undertaken in time.[2]

The set of concern areas listed in the summary above does not include the threat of nuclear war, the possibility of adverse climatic change, environmental degradation, population pressure, tyranny or malnutrition. It is not that the team is indifferent to these wider issues, but that they have deliberately chosen to leave some problems for reaearch groups such as the International Institute for Applied Systems Analysis (IIASA) which is judged to have greater resources and a more comprehensive brief.

GOALS

For some study teams the process of research itself informs and influences the goals which should be adopted. It is rather like a shopper unwilling to define precisely what is wanted before entering the shop, but waiting until the full range of merchandise has been appraised and constructing a ranking order during the course of appraisal.

FAST belongs in that other set for which the explicit statement of goals is intrinsic to the strategist's activity. A list of goals appears in "European Perspectives":

(1) The social need for *employment*: to extend employment to a sufficiently large part of the active population in order to meet legitimate aspirations and to defuse social tensions.

(2) The requirement of *external balance*: Europe must export manufactured goods so as to pay increasingly large sums for the raw materials which it is forced to import.

(3) Competitiveness requires *industrial specialisation* and increases in *productivity*, otherwise Europe will be outstripped by the US and Japan, and also caught up by certain LDC and Eastern-bloc countries.

(4) Improvement of productivity requires *technological innovation*.

(5) The need for *growth* requires concerted expansion if balance of payments problems are to be avoided.

(6) The EEC countries must accept that a *headlong flight* into *growth*, *productivity* and *interdependence* is the only route which will make it possible to satisfy the other social and political requirements.

(7) Europe should *assume its rightful place* as the leading commercial power in the world.[2]

A number of subsidiary requirements are noted—arising from difficulties which could obstruct the achievement of the primary goals. The overcoming of these difficulties could be listed as secondary goals. Some examples are listed below:

(1) greater inequalities between the beneficiaries of growth and those left behind in declining industries, skills and regions;

(2) increasing waste;

(3) damaging effects of some innovations;

(4) division of labour, mass production and automation reducing the fulfilment people achieve from their work.

TECHNIQUES

In its search for good strategy the FAST team rejects all approaches which are tinged with determinism. Hence the terms "trends, constraints, mechanisms and determinants" receive little attention and the contrasting terms "projects, desires and actors" are emphasized. Extrapolation using mathematical curve-fitting is clearly deterministic. Sectional forecasting, using a few explanatory variables of the economic, quantified type, is similarly dismissed; the argument being that economic forecasting is divorced from social and political forecasting and is itself fragmented into technological, demographic and other forecasts. Specifically it is averred that the impossibility of forecasting the future as a function solely of past data explains the weakness of classical econometric models which do not integrate qualitative and non-quantifiable parameters such as the projects and behaviour of the actors.

Historical Analysis

The FAST approach is to start with historical analysis. This consists of interpreting past trends through particular theories. It is noted that:

(1) Average rate of economic growth in 1979 was 3.3% as against 1.8% per annum during the years 1974-7, but 4.6% per annum between 1960 and 1970.

(2) There are now 6 million unemployed as against 2.1 million between 1960 and 1970.

(3) Prices rose by 14% on average between the first quarter of 1979 and the same period in 1980. The average between 1958 and 1967 was 3.3% per annum.

The question is raised: "Is the slower growth in the 1970s a harbinger of prolonged recession, or is it a necessary transition to a new system of values and another development model?" To answer the question there is a searching through the preceding historical period to disclose sources of disturbance. For example, it is observed that the Second World War redistributed political and economic power and profoundly changed the world context. Liberalization of trade and access to cheap raw materials provided the conditions for rapid growth. A wealth of other historical material is assembled in order to arrive at interpretation and recommendations:

(1) Protectionism will cause more problems that it selves. The interdependence which has resulted from the spectacular increase in international trade will amplify the negative effects of protectionist pressures.

(2) Despite the growing industrial competition from the Third World, which seems inevitable, the developed countries run ever larger surpluses in manufactured products with the developing countries.

(3) The new trend towards the growth of trade between the developing countries could help overcome the harmful effects of a recession affecting the developed countries because the developing country may develop a different economic cycle to the industrialized countries (which in fact seems the case since 1973) and thus help to stabilize the world economy.

(4) Europe, as the centre-point of trading, is an essential market for most of the other regions of the world. Its good health is vital for the world economy.

A combination of trends, correlations and theories is sued to arrive at conclusions. Thus, on specialisation:

> There seems to have been, until the crisis, a high degree of correlation between the degree of specialization of countries and the rate of growth. Should one conclude from this that intensive specialization is a condition of growth?
>
> Since the Second World War, the extension of the international division of labour - according to the Ricardian theory of comparative advantage - has contributed to the unprecendented growth of the Western countries, even if development has not often been equal, or shared by the Third World.[2]

The tools for envisaging good strategies are stated to be imagination and reflection on the future. Thus in "European Perspectives":

> Few economics or energy ministers forecast the quadrupling of the price of oil which took place in 1973-4. The models for longterm forecasting were based on a continuation of the falling price of oil in constant money terms, and thus the maintenance of growth. They also assumed a constant relationship between growth and energy consumption.
>
> But we may today be committing similar errors by using the same models to continue into the future the relationships of the past.
>
> Nor was 1973-74 the first energy crisis which the Western world experienced. In 1928 the gasification of coal and the exploitation of schists were already being seriously envisaged. The shortage of that period became a superabundance of energy with the discovery of oil in the Near East.
>
> The error was unjustified. Its cause was an absence of imagination and reflection on the future.[2]

Opinion

Among the techniques used by FAST, the role of opinion deserves comment. Consensus and Delphi methods are given short shrift. For example on energy forecasting:

> A Delphi survey would have given a fiture of 50 to 70 billion too if it had been carried out during the 1940s, and maybe double the figures in the following decade. Similar opinions do not signify coherent opinions, and in energy matters a consensus rarely makes a good forecast.[2]

Expert opinion is distinguished clearly from mere consensus. A Danzin's view is quoted that research is essential to meet the threats of the future, and to hold out the hope of a second renaissance in Europe. Frederic Saint-Geours is quoted on the current desire for decrease in public expenditure.

Surveys of opinion are considered legitimate, for example:

> Nuclear energy frightens people. Genetic control is worrying. Data processing can invade privacy and threaten jobs. But it is science that public opinion turns to in order to solve our problems. This is the overwhelming result of the recent survey carried out by Eurobarometre at the request of the Directorate-General for Research, Science and Education.[2]

SUMMARY OF STRATEGY FORMATION

The analysis and advice of the FAST team parallel closely that of consultants to a company or corporation. The company is ailing: suffering from falling profits, competition in its markets, a somewhat disaffected work force, a staff profile which shows ageing and outdated skills, wage demands, all the symptoms with which consultants become accustomed after they have examined many such patients.

The advice of FAST is the same as that proffered by company consultants: seek out the promising markets, specialize, use new technology, retrain your work force, go for growth, increase self-determination, reduce vulnerability. Watch out for demands from the workers for more participation, for the dissidents who doubt the advantages of new technology and for those who would prefer to stick with the old products and a lower turnover.

The advantage of the FAST advice is that it is robust. In the face of much uncertainty about the health of the world economy, the advice proffered remains sound through a variety of futures (given acceptance of the stated goals). This is the strength of management consultants—who have diagnosed the symptoms in former cases and observed recovery in those patients who take the medicine (also decline in those who reject their advice).

The advice, in order to be useful, must move from the general to the particular. It must be specific about specialization, preferred markets and promising areas of research and development. The FAST team set out to provide an investigation programme which could, at completion, yield detailed, particular advice.

THE INVESTIGATION PROGRAMME

The programme must meet a set of criteria:

(1) It must be acceptable to the representatives of member countries who form the advisory committee.

(2) It must be acceptable to other Directorates (apart from DG12 where FAST is sited) in the Commission so that there is no obvious duplication or overlap with existing programmes.

(3) It must relate to areas in which there are numbers of research institutions able to participate readily, so that a choice of contractor exists and there is some assurance of good-quality work being done.

(4) It must bear some relation to the background and knowledge of FAST team members.

Through a process of advocacy and discussion the areas chosen for investigation were:

(1) The Information Society.
(2) Work and employment.
(3) The "Bio-society".

The main aim of the FAST first 5-year programme is to help define Community R&D priorities with a view to the development of a consistent long-term policy on science and technology. A variety of operational means has been adopted: research by European teams, exploratory studies, state of the art reviews, individual research, seminars, and work by the FAST team itself. The involvement of a large number of experts from different disciplines and countries is well suited to the experimental nature of the programme and the complexity of the problems studied.

Work and Employment (see Tables 1 and 2)

The number of unemployed in all Community countries at the time of writing is 7 million. The FAST team sees a means of tackling the problem through the introduction and dissemination of advanced technologies with a high labour intensity, less environmental disruptive impact, low energy intensity and a high scientific content while not neglecting the large-scale capital-intensive technologies which remain necessary in some sectors and for some products.

Eleven research projects have been set up concerned with:

(1) Job creation potential from new technological changes now emerging in sectors such as chemicals, the environmental industry, construction, the repair industry and biomass energy sources.

(2) Regional employment prospects in the Europe of the 12 in the face of technological changes, sectoral readjustments and the new socioeconomic context with a view to identifying the role of suitable technological choices in securing a more balanced and more effective organization of space in Europe.

(3) Changes in attitudes towards work and the role of social innovations, which are sometimes regarded as being in opposition to technological innovations. In fact, if the crisis is to be overcome the two must be closely associated and interacting.

The Information Society (see Tables 3 and 4)

The emergence of microprocessors in the 1960s, their rapid development and the promises they held out have given greater credibility to the heralded technological and societal "revolution". What will this new information society be like? What areas will be most affected? What advantages and risks will it involve? How will they affect individuals, social groups, regions and countries?

The assessment and forecasting of employment effects are the subject of a great debate: does microelectronics represent a threat to employment or is it going to be, on balance, a new source of new job opportunities?

Seven research projects have been set up concerned with:

(1) The economic survival of European industries, bearing in mind the changes in the international division of labour which will be brought about by information technologies.

(2) The implications of an information society for individuals, groups and society in general, especially as regards representation and sharing of power, working conditions, transport and communications.

(3) The transition towards an information society. Here it will be necessary to investigate the job-creation potential of the information technologies, how to perceive and manage its benefits and risks, and how to

TABLE 1 Work and Employment - Projects and Teams[3]

Projects	Teams	Duration
Prospects for the European Chemical Industry(A2-CH)	Bureau d'Economique Théorique G Appliquée(F); Groupe d'Etude et de Recherche sur la Science de l'Université(F); in association with Centre d'Economie des Problèmes de l'Emploi et du Chômage(B)	16 months September 1980-December 1981
Prospects for the Construction Industry(A2-C)	Bureau d'Informations et de Prévisions Economiques(F)	9 months November 1980-July 1981
The Future of the Repair Industry (A2-R)	Associazione Italiana per la Ricerca Industriale(I)	18 months October 1980-March 1982
The Development of the Environmental Industry(A2-E)	Joint Unit for Research on Urban Environment(UK); in association with Institüt für Umweltschutz (D); Institut d'Urbanisme de Paris(F); Warren Spring Laboratory(UK); Geografisch en Planologisch Instituut(NL)	13 Months October 1980-November 1981
The Future of Employment in the Services(A5)	Science Policy Research Unit(UK); in association with Istituto di Studi delle Relazioni Ind.(I)	18 months October 1980-March 1982
Regional Prospects of Employment in Europe(PRESTO)(A1)	Netherlands Economic Institute (NL); in association with Association Developpement & Amenagement(F); Toegepast Natuurwetenschappelijk Onderzoek(NL); SVIMEZ(I); Centre for Environmental Studies Ltd. (UK)	18 months October 1980-March 1982
Biomass, Regions and Employment(A3)	General Technology Systems(UK); Operational Analysis Corporation (DK); Faculté Agronomique de Gembloux(B); Centre d'Economie Mathematique et d'Econométrie(B)	14 months October 1980-November 1981
Attitudes Towards Work(A6)(Seminar)	Laboratoire de Prospective et de Conjoncture(F)	Seminar to be held at end November 1981
Social Innovations in the 80's (Seminar)(A7)	Association Internationale Futuribles(F)	Seminar to be held at end May 1981
Productivity and Progress(A5)	An Foras Taluntais(Irl)	13 months October 1980-November 1981
Fiscal Policy, Social Changes and Employment	Association pour la Recherche Economique et Sociale en Europe(F)	16 months October 1980-January 1982

TABLE 2 Work and Employment - Costs and Financing
(In European units of account)[3]

Projects	Total funds	Commission contribution	Additional sources in the member states
Prospects for the European Chemical Industry(A2-CH)	140,000	40,000 (29%)	100,000 (71%) of which 27,000 Univ.de Strasbourg(F) 16,500 DGRST(F) 15,000 Serv.Prog.Pol.Scient(B) 42,000 Sources diverses
Prospects for the Construction Industry(A2-C)	27,500	20,000 (73%)	7,500 (27%) from BIPE(F)
The Future of the Repair Industry (A2-R)	72,000	45,000 (62.5%)	27,000 (37.5%) from Consiglio Naz.delle Ricerche(I)
The Development of the Environmental Industry(A2-E)	60,000	50,000 (83%)	10,000 (17%) from JURUE(UK) and INFU(D)
The Future of Employment in the Services(A8)	78,300	55,000 (70%)	23,300 (30%) from Consiglio Naz.delle Ricerche(I)
Regional Prospects of Employment in Europe (PRESTO) (A1)	325,000	200,000 (63%)	125,000 (37%) of which: 80,000 Ministère Pol.Scient. 30,000 TNO(NL) 15,000 SVIMEZ(I)
Biomass, Regions and Employment(A3)	120,000	60,000 (50%)	60,000 (50%) of which: 20,000 Serv.Prog.Pol.Scient.(B) 20,000 HIDB and SDA(UK) 20,000 3 Danish bodies
Attitudes Towards Work(A6)	52,000	30,000 (58%)	22,000 (42%) of which: 8,300 DGRST(F) 11,700 Regional authorities
Social Innovations in the 80's (Seminar)(A7)	55,000	30,000 (55%)	25,000 (45%) of which: pending official confirmation
Productivity and Progress(A5)	60,500	22,700 (37.5%)	37,800 (62.5%) of which: 32,600 An Foras Taluntais(IRL) 5,200 Irl.Found.for Human De
Fiscal Policy, Social Charges and Employment(A4)	29,000	21,000 (73%)	8,000 (27%) from DGRST(F)
Total	1,019,300	573,700 (56%)	445,600 (44%)

TABLE 3 The Information Society - Costs and Financing (In European units of account)[3]

Projects	Teams	Duration
Microelectronics innovations in the context of the international division of labour(B1)	Battelle-Institut CV(D)	21 months September 1980-May 1982
Job creation potential of the information technologies(B2)	SEMA(F); in association with International Systems Corporation of Lancaster(UK); IRIS(F)	19 months September 1980-March 1982
Long-term options and forecasts for transport in Europe(B3)	Institute for Transport Studies University of Leeds(UK)	15 months November 1980-January 1982
Representation and sharing of power in an information society(B4)	Bureau d'Informations et Prévisions économiques (F); Jens Bisballe Plan Laegning(UK); in association with Telecommunications Directorate Ltd(UK)	16 months November 1980-February 1982
The impact of information technologies on the way of life: work, leisure, transport and communications(B5)	Centre de Recherche sur le bien-être(F); Laboratoire d'Economie des Transports, LYON(F); National Board for Science and Technology (IRL)	17 months October 1980-February 1982
Distribution of benefits and risks associated with microelectronics applications (Seminar)(B6)	Manchester Business School(UK)	Seminar to be held in January 1982
Man-machine relations: dangers and remedies(B7)	Machine Intelligence Research Unit(UK)	12 months November 1980-October 1981, plus seminar in February 1982

negotiate the changes.

The 'Bio-Society' (see Tables 5 and 6

There is growing interest in the prospects opened up by new scientific and technical discoveries in the field of biotechnology for the production of new goods and services.

In the fields of food production, health, chemistry, energy and environment there are various potential applications: for example, biomass production via bacteria, yeasts, fungi, and algae for food production; *in vitro* hormone production; use of immobilized micro-organisms and enzymes to catalyse

TABLE 4 The Information Society—Costs and Financing (in European units of account)[3]

Projects	Total funds	Commission contribution	Addition sources in the member states
Microelectronics innovations in the context of the context of the international division of labour(B1)	285,500 (100%)	190.000 (66.5%)	95,500 (33.5%) from: different industrial companies
Job creation potential of the information technologies(B2)	325,000 (100%)	180,000 (55%)	145,000 (45%) of which: 34,000 DGRST(F) 29,000 SEMA(F) 33,000 Man.Serv.Comm.Dept. Envir.(UK) 33,000 Midland Bank(UK) 16,000 Other French bodies
Long-term options and forecasts for transport in Europe(B3)	47,500 (100%)	25,000 (53%)	22,500 (47%) from: Univ.of Leeds(UK)
Representation and sharing of power in an information society(B4)	178,000 (100%)	95,000 (53%)	83,000 (47%) of which: 29,000 Ministère de l'Ind. 14,000 Sec.de la Recherche 23,000 Dir.Télé.,LTP(D) 17,000 EDG(F)
The impact of information technologies on the way of life: work, leisure, transport and communications(B5)	214,000 (100%)	100,000 (47%)	114,000 (53%) of which: 59,000 NBST (IRL) 55,000 CNRS(F)
Distribution of benefits and risks associated with microelectronics applications (Seminar)(B6)	72,000 (100%)	45,000 (62.5%)	27,000 (37.5%) of which: 15,000 IBM(UK) 8,000 Cable & Wireless Co. 4,000 N.Holland Publ.(NL)
Man-machine relations: dangers and remedies(B7)	30,000 (100%)	15,000 (50%)	15,000 (50%) from Univ.of Edinburgh(UK)
Total	1,152,000 (100%)	650,000 (56%)	502,000 (44%)

organic reactions; cultivation of plants producing hydrocarbons and resins; and the purification of municipal and agricultural liquid effluents.

Twelve research projects have been set up concerned with:

(1) The definition of a European strategy based on the systematic gathering of available information and identification of the opportunities and priorities for research and development.

TABLE 5 The Bio-Society - Projects and Teams[3]

Projects	Teams	Duration
A Community strategy for European biotechnology (Symposium-seminar)(C1-1)	Dechema(D) for the European Federation of Biotechnology	Symposium to be held in September 1981
Biotechnologies in the production of chemical feedstocks and derived products(C1-2)	International Research and Development Co. Ltd.(UK)	19 months September 1980-March 1982
Proteins: probable impacts and development strategies for biotechnology in European agro-food systems(C1-3)	Association Promotech (Institut National Polytechnique de Lorraine)(F)	18 months September 1980-February 1982
Community strategy for biotechnology: analysis of scientific disciplines germane to biotechnology (C1-4)	Technology policy unit (UK) in cooperation with Gorlaeus Laboratory, Chemistry and Society Programme(NL)	18 months October 1980-March 1980
Information technologies in the development of biotechnology (bio-informatics)(C1-5)	Negotiations in progress	
Manpower and training implications of the expansion of biotechnology-based industries (symposium-seminar)(C2)	Société de Chimie Industrielle(F), for the European Federation of Biotechnology	12 months January 1981-December 1981 Seminar to be held October '81 prov.
Social acceptability of biotechnologies(C3); Social and political choices that may be created by progress in biotechnologies(C6)	Negotiations in progress	
Proteins: implications of biotechnologies for developing countries and international trade(C4-1)	Centre de Recherche en Gestion Internationale (B)	12 months November 1980-October 1981
Implications for the Third World of biotechnology (C4-2)	Conservatoire National des Arts et Métiers, Centre Science-Technologie-Société(F)	12 months October 1980-September 1981.Seminar to be held Sept. 1981
Technological forecasts of downstream processing in biotechnology(C5-1)	Univ.of Manchester Institute for Science and Technology(UK)	24 months July 1980-June 1982
Detoxification—as therapeutic techniques, problems and prospects(C5-2)	Assoreni(I)	8 months October 1980-May 1981
Environmental biotechnology: future prospects (Seminar)(C7)	National Board for Science and Technology(IRL)	Seminar to be held in April 1982

TABLE 6 The Bio-Society - Costs and Financing (in European units of account)[3]

Projects	Total funds	Commission contribution	Addition sources in the member states
A Community strategy for European biotechnology (symposium-seminar) (C1-1)	90,000 (100%)	90,000 (100%)	Voluntary efforts from the working groups of the European Federation of Biotechnology has been promised but not included in the figures
Biotechnologies in the production of chemical feedstocks and derived products (C1-2)	49,000 (100%)	49,000 (100%)	Cross-support from contractor parallel industrial studies
Proteins: probable impacts and development strategies for biotechnology in European agro-food systems (C1-3)	72,000 (100%)	36,000 (50%)	36,000 (50%) from Promotech (INPL) (F)
Community strategy for bio-technology: analysis of scientific disciplines germane to biotechnology (C1-4)	39,000 (100%)	19,500 (50%)	19,500 (50%) from the Univ. of Aston (UK) and Leiden (NL)
Information technologies in the development of biotechnology (bio-informatics) (C1-5)	Under discussion		
Manpower and training implications of the expansion of biotechnology based industries (symposium-seminar) (C2)	30,000 (100%)	30,000 (100%)	Voluntary efforts have been promised by the institutes in the Education working party of the European Federation of Bio-technology but are not included in the figures.
Social acceptability of bio-technologies (C3); Social and political choices that may be created by progress in bio-technologies (C6)	Under discussion		
Proteins: implications of biotechnologies for developing countries and international trade (C4-1)	53,000 (100%)	22,000 (42%)	31,000 (58%) of which: 22,000 Serv. Prog. Pol. Scient. (B) 9,000 CRGI Univ. Louvain-la-Neuve (B)
Implications for the Third World of biotechnology (C4-2)	40,000 (100%)	29,000 (73%)	11,000 (27%) of which: 8,500 CNRS (F) awaiting confirmation

TABLE 6 The Bio-Society - Costs and Financing (in European units of account)[3] (continued)

Projects	Total funds	Commission contribution	Addition sources in the member states
Technological forecasts of downstream processing biotechnology (C5-1)	156,000 (100%)	56,000 (36%)	100,000 (64%) from UMIST on top of contributions from SRC(UK) and British chemical firms
Detoxification—as therapeutic technique: problems and prospects (C5-2)	22,000 (100%)	7,000 (32%)	15,000 (68%) from le groupe ENI(I)
Environmental biotechnology: future prospects (Seminar) (C7)	54,000 (100%)	27,000 (50%)	27,000 (50%) from NBST (IRL)
Total	605,000 (100%)	365,500 (60%)	239,500 (40%)

(2) The manpower and training implications of the expansion of biotechnology in different industrial sectors.

(3) The study of the effects that the development of the new technologies may have on our trading and industrial relations with the Third World (reductions in raw material imports but also increased opportunities for industrial trade).

(4) Examination of the social implications of biotechnology in defining R&D priorities, in order to promote its socially harmonious development.

PROGRAMME SCALE

The programme of research is distributed among 49 contract research teams from Community countries at an overall cost of 2.8 million EUA. The FAST final report will be written in 1982. This report will distil the implications for European science and technology that emerge from the programme.

NOTES

1. Kennet, Wayland (ed.): *The futures of europe*. Cambridge University Press, Cambridge, 1976.

2. Godet, and Ruysser, *European Perspectivies - The Old World and the New Technologies*. Office for Official Publications of the European Communities, 1980.

3. Commission of the European Communities, DG XII Research, Science and Education: FAST Research Activities EUR 7102. Office for Official Publications of the European Communities. Catalogue number: CD-NW-80-006, 1980.

Part III

Modern Trends in Scientific and Technological Development and Its Impact on Man and Society

FHN - E*

The future of robotics: a meta-review

R. U. Ayres

INTRODUCTION TO ROBOTICS

What is a Robot?

Industrial robots are machine tools. They are not human-like androids that can walk around and converse, as in literature and films. More realistically, they are programmable manipulators which can move parts or tools through a prespecified sequence of motions. But they are also—at present—immobile, clumsy, blind (for the most part) and very stupid. They usually have one arm and two fingers with a total of 4, 5, 6 or 7 degrees of freedom. Reprogrammability means that the robot's actions can be modified by changing control settings, without changing the hardware. They combine some attributes of traditional machine tools as well as attributes of machine tool operators. Like a machine tool, the robot can repeat the same task for prolonged periods with great precision. Like an operator, it is flexible enough to be taught to do a new task, and it can use accessory tools to extend its range of physical capabilities. One classification of robots is shown in Table 1. Note that the first two categories in the Japan Industrial Robot Association (JIRA) classification are not defined as robots in the United States.

Robots are valued in industry for the usual qualities of machines: untiring availability, predictability, reliability, precision and (relative) imperviousness to hostile environments. They do not, as yet, possess several important capabilities which come naturally to humans: the ability to react to unforeseen circumstances or changing environments, and the ability to improve performance based on prior experience. State-of-the-art robots (mostly in research labs) do have crude senses of "sight" and "touch", and limited capability to coordinate their manipulators with sensory input. Because of current limitations, today's robots are usefully employed in highly structured industrial environments where practically all of the variability and decision-making can be engineered out of the workplace. Existing uses of industrial robots all involve repetitive preprogrammable tasks such as spot welding, spray painting, palletizing, and the loading and unloading of many types of metal-forming and metal-cutting machines. The next generation of sensory-based robots will be able to perform a broader range of tasks under less structured conditions, in addition to becoming cheaper and easier to use. Expected uses of robots with vision and improved feedback control will include inspection, assembly, heat treatment, grinding and buffing, and electroplating. Capabilities of commercially available robots, and capabilities under development for future robots, are listed in Table 2.

TABLE 1 Classification of Input Information and Teaching Method (JIRA)

1. Manual manipulator

   A manipulator that is directly operated by a man.

2. Fixed-sequence ("pick and place") robot

   A manipulator which performs a sequence of steps specified by means of cams, mechanically set limit switches and/or hydraulic or pneumatic values.

3. Variable sequence robot

   A manipulator as defined above, for which the instructions can be specified by resetting electrical connections.

4. "Playback" robot

   A manipulator that can remember and repeat any operation after being instructed by an operator; subcategories are known as "walk through" and "lead through".

5. Numerically controlled (NC) robot

   A manipulator that can receive instructions in digital form via a magnetic tape or directly from a computer.

6. "Intelligent" robot

   A robot that can modify its own actions in response to data obtained through its sensing and recognitive abilities.

Eventually, many of the "hands on" tasks performed by production workers on the factory floor will be done by robots in computer controlled manufacturing systems. Programmable automation is beginning to replace the current generation of manually controlled machines. This transition will undoubtedly continue for many decades. There is a potential for significantly improving the productivity of the manufacturing sector, and increasing the wealth-producing potential of the economy as a whole. We also face significant social impacts, such as the short-term prospects of technological displacement, and the longer-term prospects of basic structural shifts in the economy.

Some Historical Background

The term "robot" allegedly stems from the Czech word "robotnik", meaning serf or worker. It was first introduced into the popular language by Karel Capek, a Czech playwright, in *RUR*. The concept of programmable machinery, however, dates back much earlier, to the eighteenth century, when the Frenchmen Bouchon Falcon de Vaucanson and Jacquard developed variable sequence mechanical looms. Jacquard's looms were controlled by punched cards. Spencer's "Automat", a cam programmable lathe used for producing screws, nuts, and gears in the 1870s, was controlled by fitting guides to the end of a rotating drum. Since then, mechanical controls have proliferated in the machine tool industry.

TABLE 2 Robot Capabilities

	Commercially available capabilities (1980)	Capabilities sought for the future
Learning	On-line programming via teach/playback modes; teaching in multiple coordinates; local and library memories of any size;	General-purpose robot programming languages; off-line programming; learning with experience;
Decision-Making	Program selection by random stimuli	World model of working environment
Sensing	Computer interpretation of sensory data; computer interfacing; 2-D vision with binary recognition; force/torque sensing; limited speech input	Positional sensing; 3-D vision with grey levels and colour; tactile sensing; voice communication; improved processing of sensory inputs; coordination of multiple sensory inputs and control
Manipulation	Six infinitely controllable articulations between base and gripper; point-to-point control; continuous path control; position accuracy repeatable to 0.3 mm; handles up to 150 kg	Miniature manipulators; greater positional accuracy; greater dynamic control; general-purpose hands; multiple hand-to-hand coordination
Mobility	Synchronization with moving workpieces	Programmable omni-directional mobile bases; self-navigating mobile bases; "walking" robots
Reliability	400 hours for mean time between failure failures	Self-diagnostic fault tracing

Mechanical manipulators have an even longer history. Articulated limbs for amputees date back to at least 1509. The military surgeon Paré was providing metal arms with flexing elbows and hands that could open and close, by the late sixteenth century. In 1892 Seward Babbitt of Pittsburgh patented a rotary crane with a motorized gripper for removing ingots out of furnaces. The first jointed mechanical arm which could "play back" a series of motions was developed by Pollard in 1938. The machine was specialized for spray painting. The first general-purpose playback unit for controlling machines was developed by George Devol in 1946. He licensed the device to the Remington Rand Corporation, which intended to use it for the Univac Computer, which had just been developed. The controller was not fast enough for the desired purpose, and the patent was returned to Devol. In 1954, Devol developed the first general manipulator with a playback memory and point-to-point control. The patent for this Programmed Article Transfer was issued in 1961. The patent states: "Universal Automation, or Unimation, is a term that may well characterize the general object of this invention." Devol's early patents were later sold to Condec Corp., and formed the basis of Condec's robot division, Unimation, Inc., the world's leading robot manufacturer today. In the period between 1954 and 1963, Devol and several others patented the major features of the first generation of robots. Robots were commercialized in the United States in 1960 by several firms. Unimation concluded a licensing agreement with Kawasaki in Japan in 1968. Kawasaki is currently the leading Japanese producer of robots.

Early robots had computer-like functions, such as memory, but were made up of electronic logic components "hard-wired" to perform a specific set of tasks. Electronic controls were used essentially to duplicate the functions of other "hard automated" control functions. Robots controlled by general-purpose computers were developed in the early 1970s. The first mini-computer controlled robot was commercialized in 1974 by Cincinnati Milacron Corp., a large machine tool manufacturer. Microprocessor controlled robots followed soon thereafter. The computer controlled, or "soft-wired" robot, is far more powerful than a machine with specialized electronic logic circuits. It can work in several coordinate systems, be programmed "off-line", interface with sensors, and so on. Computer controlled robots are now evolving into specialized peripheral features of a general-purpose computer, that can also undertake broader production management functions.

Recent History

Industrial robots came into use in the United States during the 1960s but there were probably only a few hundred in use by 1970. Despite yearly predictions by robotics manufacturers that the technology was about to "take off", it probably took about 7 years for the number to double. The various estimates shown in Figure 1 suggest that the US had about 1300 robots in 1974 and perhaps 2500 by 1977. But industry growth slowed, at least temporarily. In 1980 it appears that there were approximately 3500 industrial robots in use in the US, about 7500 in Japan, perhaps 4000 in Europe, 2000 in the USSR and a world total of about 18,000 (based on the US definition of robot).

Stories in the press also have reported that the number of orders placed in the past year, in the US about equals the number of robots already in use. Industry and trade journals over the last year have consistently reported an "explosion" in orders for industrial robots. General Motors, with 300 robots installed as of January 1981, has announced plans to increase that number to 15,000 by 1990. General Electric Corp. has recently entered the robotics field as a manufacturer (and is its own largest customer). Three large computer firms—IBM, Texas Instruments, and Digital Equipment—are widely regarded as potential new entries to the field (*Business Week*, 17 August 1981), as is the other electrical equipment giant, Westinghouse Electric Corp. Bendix Corp., the number two manufacturer of machine tools in the United States (following its acquisition of Warner and Swasey) is also now manufacturing robots. New robotics firms or divisions of existing firms are being formed almost monthly.

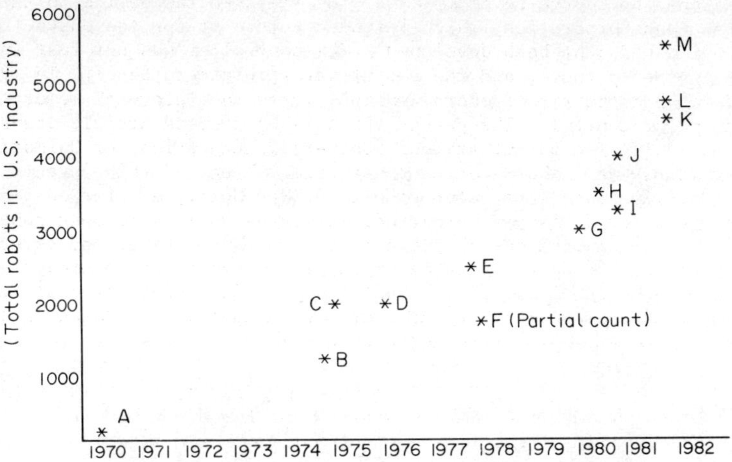

Point	No. of Robots	Date	Source
A	200	1970 (April)	Engelberger, First National Symposium on Industrial Robots, 1970
B	1200	1974 (Dec.)	Frost and Sullivan, US Industrial Robot Market, 1974
C	2000	1975 (Dec.)	Frost and Sullivan, The Industrial Robot Market in Europe, 1975
D	2000	1976 (Dec.)	Eikonix Technology Assessment, 1979
E	2400	1977 (Dec.)	Eikonix Technology Assessment, 1979
F	1600	1978 (Dec.)	American Machinist 12th Inventory, 1978
G	3000	1980 (Jan.)	Walt Weisel, Prab Conveyors
H	3500	1980 (June)	Business Week, verified by Cincinnati Milacron
I	3200	1980 (Dec.)	General Motors Technical Staff, (Bache, Shields estimate)
J	4000	1980 (Dec.)	Walt Weisel, Prab Conveyors
K	4370	1981 (Dec.)	Paul Aron, Daiwa Securities
L	4700	1981 (Dec.)	Robot Institute of America Survey
M	5500	1981 (Dec.)	Seiko Inc. Marketing Dept.

Figure 1. Estimates of the robot population in the United States, 1970-80.

Despite an 8-year US head start over Japan, which produced its first robot in 1969 (based on a US licence) Japan is now considerably ahead of the US in both production and use. Japanese production in 1980 was 3200 units, worth $180 million as compared to 1269 units ($100 million) in the US. There are nearly 140 robot producers in Japan.

The USSR did not produce its first robot until 1971-2. Yet, according to D. Tesar (University of Florida) there are now 2000-3000 units in place. According to *Business Week* (17 August 1981) there are already 70 different models in production and 200 experimental units are said to have sensory feedback capability.

Applications of Industrial Robots

Robots are mostly used today in manufacturing operations in batch production, especially of metal and plastic products. Robots are appropriate where the operation is repetitive but where the motions must be changed (i.e. reprogrammed) at fairly regular intervals.

Robots are not really cost-effective in most "custom" manufacturing applications because in such cases a large fraction of the labour time is spent setting up the machines. This still requires the active involvement of a skilled machinist. For one-of-a-kind and prototype products it is easier for a skilled machinist to make the piece than to figure out how to teach a robot to do it.

Robots are not generally cost-effective in mass-production applications either, because specialized mass-production machinery can usually perform the operations more efficiently. Mass-production machinery, or "hard automation", is ultra-specialized to repeat a fixed sequence of operations at high speeds for very long periods of time. Automobile engines and transmissions are manufactured in this way. It is always difficult and expensive, if not impossible, to configure the hard automated system for another product. It is usually cheaper to scrap the machinery, and rebuild the system from scratch.

Specific tasks for which robots are used today are shown in Table 3. Experimental and future applications are also suggested. A summary characterization of benefits is shown in Table 4.

TABLE 3 Robot Application Areas

Commercially available robots

Die casting	Plastic moulding
Spot welding	Foundry work
Arc welding	Machine tool loading
Investment casting	Heat treatment
Forging	Deburring of metal parts
Press work	Palletizing
Spray painting	Brick making
Glass manufacturing	Dimensional inspection

Experimental robots

Assembly
- electromechanical assembly
- insertion of other small, fragile components

Inspection
- of electromechanical parts and assemblies
- of electronics parts and assemblies

Sheep shearing

Future robots

Exploration and extraction of materials
- in subsurface mines
- undersea
- in outer space

Space manufacturing

TABLE 4 Characterization of Robot Benefits

	Current industrial robots	Experimental industrial robots	Future robots
Areas of application	Highly structured manufacturing task	Variable manufacturing task	Subsurface, undersea, space
Description	General-purpose machine tool with no sensing capabilities	General-purpose machine tool with limited sensing and decision-making	Autonomous system guided by artificial intelligence
Benefits	Factor saving capital material labour	Factor saving capital material	Resource extending

FUTURE ROBOT CAPABILITIES: METHODS AND PROBLEMS

There are two basic approaches to forecasting, each of which has its advocates. One approach focuses on the driving forces behind technological change and starts from the question: "What new technological capabilities are needed?". The other approach starts from the existing situation (as perceived by the technologists in the field) and raises the question: "What new capabilities are possible (likely)?". There is, of course, an interplay between these approaches. For instance, the historical rate of progress of a given technology certainly reflects, to some extent, the amount of attention and resources it has received. Progress in computer technology in the last several decades unquestionably reflects heavy investment in the field. Obviously, attention and resources are attracted, in turn, because of perceived societal needs (or, at least, demands) for the capability being developed.

In the case of robots the primary driving force for developments to date has been economic. In Japan, rapid growth has created a labour shortage which in turn induces Japanese industry—with full cooperation of the workers—to implement new forms of mechanization and automation as fast as they are perfected. In the US and Europe growth has been slower and labour has not been scarce in the past decade. In addition, several factors have combined to reduce the availability of investment capital. Thus robot utilization has been slower. It has accelerated recently in the automobile industry because the market has shifted dramatically towards smaller cars, leaving the US firms in a very bad competitive situation *vis a vis* the Japanese. Robots are now being installed as rapidly as possible, mainly for welding and spray painting, both to cut labour costs and to improve product quality. The major automobile firms, as well as the large electrical equipment and farm equipment manufacturers, are also anxious to use robots for computer controlled machining centres and for parts assembly and inspection.

Demand for robots capable of doing parts assembly tasks provides the motivation for current research in robotics. In particular, the objective of current R&D is a *sensor-based robot*. Most robots that are on the market today have closed-loop feedback control, such that joint position is continuously compared to a programmed trajectory and the controller tries to minimize errors. But mechanical systems with closed-loop controls are inherently not precise enough to carry out most assembly and inspection tasks. External sensory information (e.g. vision or force feedback) is needed to achieve the necessary accuracy. To provide this capability is the chief objective of most advanced robotics research at present.

Having said this, it would be desirable to define quantitative measures of robot capabilities, to monitor the recent past rate of progress and, by extrapolation, to permit intelligent estimates of when key future milestones may be achieved. Planners of future manufacturing facilities, for instance, must decide a number of years in advance, whether to design an assembly line to be "manned" by humans or robots, for example.

The first step in developing a forecast is taxonomic: it is to identify relevant and quantifiable performance measures. A review of the robotic literature suggests the following list:

mobility (of the robot as a whole);
degrees of freedom of the arm and "end effector" (up to 6);
arm speed;
workspace volume;
payload weight;
effector flexibility (ability to handle generalized shapes, etc.);
positioning accuracy at specified points;
repeatability of positioning;
number of points that can be specified along a path (i.e. approximation to continuous path);
reliability (mean time to failure);
down-time for program change;
time required for program change (off-line)

Note that while robots may be classified according to whether they possess external sensory feedback (e.g. vision), such capabilities are merely contributory to a primary performance criterion, viz. positioning accuracy.

Many of the above measures are readily available from manufacturers' literature, but others are not. Effector flexibility, for instance, is difficult to define precisely. It will probably have to be further broken down into component measures. In other cases the *desired* measure would have to be constructed from available measures. For instance, the working volume can be computed from a knowledge of the axial, radial and angular ranges of the arm. The number of points that can be specified along a path may only be inferred from the size of the memory of the controlling microprocessor or mini-computer. Down-time for program change would, of course, depend on the kind of change—though comparisons could easily be made between different robots.

The above list already reflects a number of successive approximations to a satisfactory taxonomy. Very probably it will continue to evolve.

The next step, in principle, would be to combine the various measures of performance to obtain a *single* overall measure, e.g. of robot sophistication. The simplest scheme would be to construct a linear combination of individual measures, using coefficients representing "relative importance". In fact, there are several alternative approaches to constructing aggregate measures depending on what supplementary data is available.

The so-called "hedonic" variable method is to regress market prices against technical characteristics, for each model of robot.[1] We do not have sufficient data to do this, unfortunately. Another possible approach proposed by Sahal[2] starts from a set of observations, or technical descriptions of robots available in two different years (say 1965, 1980) and constructs a statistical function—the discriminant—which maximizes the probability that any new observation would be correctly placed with its generation. This function can be interpreted as a measure of (technological) "distance" between the two sample years—hence an aggregate measure of technology *per se*. This method was considered, but not adopted.

A third approach, which we tried out on an experimental basis, was based on work by T.L. Saaty.[3] The weights for an aggregate measure were determined by a procedure that uses pairwise comparisons of relative desirability (or utility) based on interviews of a number of experts at Carnegie-Mellon University and Westinghouse. The hierarchy used in the experiment is shown in Figure 2. At each level of the hierarchy relative desirability is based on a consistency-weighted average of expert opinions. In effect, each expert fills in the upper off-diagonal elements of a matrix m_{ij} by attaching a numerical value to the ratio of importance of the ith attribute relative to the jth. Diagonal elements are unified by definition and lower off-diagonals are defined as reciprocals viz, $M_{ji} = M_{ij}^{-1}$. The principal eigenvector of the matrix (normalized) becomes the "preferred" vector of weights for the attributes, and the difference between the trace of the matrix and the sum of the eigenvalues is a measure of the internal inconsistency of the respondent. (A matrix whose pairwise comparisons were all perfectly consistent with each other would yield an inconsistency measure of zero.)

Calculated attribute weights are shown in Table 5. We used these numbers, along with technical data, to compare four robot models in terms of overall relative utility. We found, however, that the results did *not* correlate well with actual prices. This raises serious doubts about either the interview procedure or the initial hierarchical classification scheme (Figure 2). It is possible that we will repeat this exercise with a modified hierarchy and a larger number of interviews next year.

We are currently planning to test still another variant of the "demand driven" approach to the problem of forecasting robot capabilities. It can be argued that, for the special case of robot vision, at least, the "pacing" parameter is computer capability. At present the capabilities of solid-state cameras are far inferior to the human retina, but the biggest difference between human and machine vision systems is in the data processing area. It is estimated that the human visual cortex—roughly 10% of the brain—can handle 100,000 times more data per second than a typical minicomputer (such as the PDP-11).

Experience over the last 30 years has shown that computer "power" (per unit cost) has consistently increased by a factor of 10 every 5 years, or by a factor of 10^6 since 1951 when the first commercial Univac computer was built. Most experts today believe this rate of progress will continue for at least the next 20 years; it seems to follow that robot vision capabilities are likely to improve very rapidly. Ten years from now robots should have vision capability adequate for most parts assembly and inspection tasks, for instance.

By an obvious extension of the above argument, the "intelligence" of robots, should also be dramatically improved. Intelligence, in this context, can be defined as an ability to solve "ill-structured problems"[4]—exactly the sort

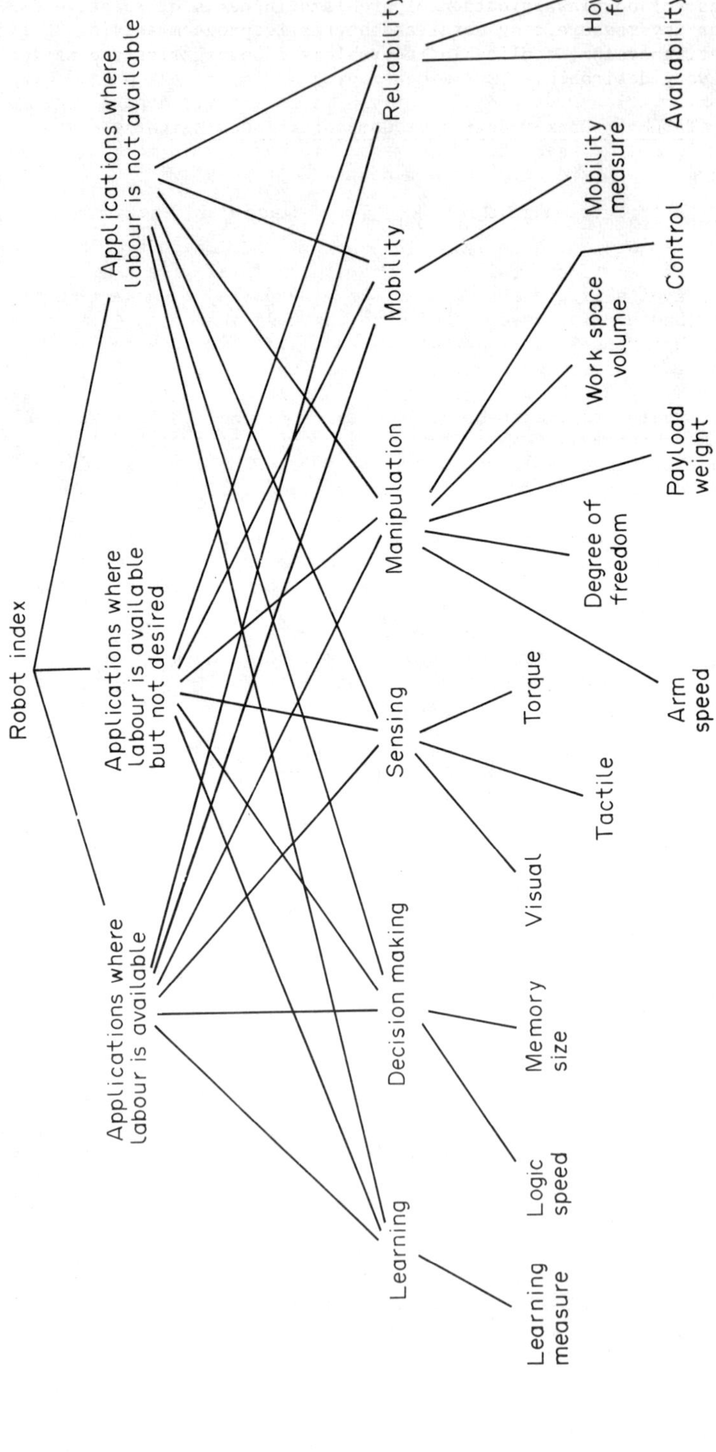

Figure 2. Composite robot index hierarchy.

of problems a robot must solve in trying to perform a task in a variable environment. Unfortunately, this observation is essentially qualitative. We have as yet no classification of problems in terms of relative difficulty, nor any measure of generalized heuristic problem-solving ability.[5] Such a rating system, applicable to problems of navigation and manipulation, would be very desirable.

TABLE 5 Rank Ordering of Composite Index Weights

Attribute measure	Consistency corrected composite weight
Availability	0.330
Learning	0.124
Memory size	0.120
Visual sensing	0.097
How it fails	0.060
Logic speed	0.053
Mobility	0.042
Tactile sensing	0.037
Number of degrees of freedom	0.030
Positioning accuracy	0.026
Torque sensing	0.021
Maximum arm speed	0.020
Maximum payload weight	0.016
Path control	0.016
Working space volume	0.010

Source: G. Miller, unpublished

FUTURE ROBOT UTILIZATION: METHODS AND PROBLEMS

The term "utilization" as used in the title of this section goes beyond the notion of aggregate demand, to cover distributional issues. That is, we ask "what specific jobs (tasks) will robots be able to do, and in what industries, states and cities are these jobs now located?"

It is worth mentioning at the outset that very detailed job classifications and industry classifications have evolved over the past several decades and the US Bureau of Labor Statistics has compiled data on the number of workers by job type, in each major industrial sector. More detailed data (for instance, on regional location of jobs) exist in computers in Washington, DC, although publication is restricted because it might tend to reveal information about individual employers, which is prohibited under the law governing the activities of the Census.

In principle we would like to analyse each job type in terms of its suitability for robots. A case-by-case analysis is impractical, however, because of the very large number of job descriptions in a modern industrial economy. The US Department of Labor has compiled a bulky "Dictionary of Job Classifications" consisting of short descriptions of each job type. However there are many thousand such descriptions, and the published descriptions do not contain enough information to permit an unequivocal decision as to whether a robot of specified capability could perform the job. They do, however, include a numerical scale of "skill" required, from lowest to highest:

(1) handling of materials (e.g. lifting, shifting, dumping)
(2) feeding/off-bearing (loading/unloading workpieces)
(3) tending machines (on/off, watching feed, etc.)
(4) manipulating (e.g. assembly/disassembly)
(5) driving-operating (e.g. a vehicle)
(6) operating-controlling (monitoring dials, etc.)
(7) precision-working
(8) "setting up" machines for new jobs.

In principle, one could imagine relating potential for robot-substitutability to the level of skills required on the job. Unfortunately, however, robot capabilities do not correspond neatly to human skills. Furthermore, it is generally found that robots do not fit exactly into existing job "niches". Much more often than not, the work must be reorganized to give robots the repetitive tasks they are capable of doing, while retaining humans for jobs requiring more judgment or better eye-hand coordination.

We were forced, ultimately, to fall back on the use of a survey. With help from the Robot Institute of America (a trade association of robot manufacturers, distributors and users) we sent a questionnaire to all members (firms) asking: "What fraction of the jobs in each category could a robot—of given capability—perform?" This strategy assured us, at least, that the respondent already knew something about robot capabilities. All of the 16 firms that ultimately provided detailed responses were already robot users or were actively exploring the possibility. A Level I robot was defined as any robot *without* sensory feedback or "intelligence" (i.e. categories 3-5 in Table 1) while a Level II robot corresponds to category 6 of Table 1.

The responses were compiled and weighted by the size of the responding firm. Overall estimates of jobs that might be done by robots were obtained by multiplying resulting percentages by total employment in each job category. Results are shown in Table 6.

The foregoing analysis is obviously deficient in terms of forecasting the rate of substitution of robots for human workers. This subject is of particular interest to manufacturers of robots and can be called a "market forecast". A number of such forecasts have been made by various individuals and groups, largely on the basis of extrapolating recent market growth rates. The Japan Industrial Robot Association (JIRA) expects robot use in Japan to grow at a rate of 50% per year for the next 5 years, and *Business Week* magazine quotes Soviet planners to the same effect. In the US, IRA projections have been slightly more modest, in the 35% per annum range based on a more restricted classification of robots.

We are trying to address this question of market growth in a more sophisticated and quantitative fashion. The viability of technologically advanced (seeing, feeling) robots depends on their cost-effectiveness. Capability alone will not assure widespread use. The critical question for a potential user is whether the capability purchased with a robot is economically competitive with either human labour, on the one hand, or "hard automation" on the other. Stating the issue in still another way, we need to address the future demand for sophisticated robots as a function of price.

It can be reasonably assumed that in a young competitive industry, prices in the long run will stay fairly close to production costs. This has been the

TABLE 6 Estimates of the Number of Jobs Robots Could Do in All Manufacturing

Occupation	Total employed 1979[a] All manufacturing	Percentage of jobs which could be done by:[b] Level I	Level II	Number of jobs which could be done by: Level I	Level II
Assemblers	1,289,000	10	30	128,900	386,700
Checkers, examiners[c] and inspectors	746,000	15	35	111,900	261,100
Packers and wrappers (except meat and produce)	626,000	15	40	93,900	250,400
Production painters	185,000	45	65	83,250	120,250
Welders and flame-cutters	713,000	25	50	178,250	356,600
Machinists and machine operators[d]	3,027,000	20	50	605,400	1,513,500
	6,586,000[e]			1,201,600	2,888,450

a Employment figures from Statistical Abstract, 1980, Table 697, Persons Employed by Sex, Race, and Occupation, 1972 and 1979.

b The percentage of jobs which could be done by Level I and Level II robots are based on survey results within the metal-working sector. Percentage estimates are the average weighted responses of survey results, rounded to nearest multiple of 5.

c Includes testers and inspectors. Since there are more than four times as many testers as inspectors in metal-working, the percentage estimates for inspectors are used.

d Includes the following occupational categories: machinists and job-setters, precision machine operators, machine operators, and punch press operators.

e The difference between this figure and the estimate of the number of operatives given in the table "Estimates of Percent Coverage in Metal-working for Selected Occupations" is as follows:

6,586,000	−	642,000	+	780,000	+	263,000	=	6,987,000
(total given above)	−	(machinist and job-setters)	+	(freight and material handlers)	+	(cutting operatives, nec)	=	(total operatives and unskilled workers)

recent pattern in the microelectronics industry which, of course, will provide some essential components of the sophisticated robots. Costs, on the other hand, are likely to be a function of cumulative production experience, which usually gives the market leader a competitive edge. As a rough generalization, new market entries can only achieve a significant market position through "buying" market share (by producing at a loss) or by exploiting a technological superiority. Some small new entrants to the field are likely to succeed by the second strategy, while one or two major firms—such as IBM, Texas Instruments or General Electric—might achieve dominant positions via the first route.

It is vital to recognize, of course, that the market is a very dynamic one. Sales and production are currently rising rapidly and the "experience curve" suggests that prices—in real terms—should begin to decline fairly sharply. As a rough guide, when the number of robots in service doubles (probably within 2 or 3 years) production costs would be expected to decline by 20-30%. Based on this relationship, known as the "experience curve", some fairly spectacular price reductions might be expected. For instance, assuming a modest 20% decline with each doubling of cumulative production (i.e. experience) and a 35% annual market growth, the manufacturing cost "index" of robots would drop from 100 in 1980 to 52 in 1985. If the market grew faster (50% p.a.) the cost index would drop to 46. If the rate of price decline were 30% per doubling of experience, the 35% p.a. growth rate would result in a 1985 cost index of only 35, while the 50% p.a. growth rate would result in a cost index of only 29.5 by 1985.[6] Roughly, a system costing $100,000 in 1980 could cost anywhere from $30,000 to $50,000 by 1985 (in 1980$). For the next few years, as demand grows faster than output, prices may not decline as fast as costs and profit margins may rise, attracting more new entrants to the market. Inevitably, however, a "crunch" will occur in a few years when supply catches up with demand and market prices would drop sharply, bringing prices into line with costs.

When declining prices follow declining costs, demand for robots will continue to grow. Wage costs for human labour can also be expected to continue to grow, much as it has in the recent past. The passing of the "baby bulge" means that fewer persons will enter the labour force in the 1980s than in the 1970s. This will continue to drive up the cost of human workers. Thus, the basic economic justification for robot use is straightforward and the expected value of benefits will grow rapidly over the next two decades.

It can be shown that, with a rapid fall in prices, the discounted present value of purchasing a robot may also decline over time. This could give the would-be purchaser an incentive to wait! This is, of course, paradoxical and it seems reasonable to assume that the market will *not* grow (hence prices will *not* decline) so fast as to induce such delays.

With this further assumption, a maximum market growth rate can be calculated as a function of the initial capital payback period, "experience" parameter, and discount rate. The full derivation is omitted, but results are summarized in Table 7. Note that the results are extremely sensitive to the initial payback period and discount rate. High rates of market growth (∿35% p.a.) are inconsistent with high discount rates and initial payback periods longer than a few years.

SOCIOECONOMIC IMPACT OF ROBOTS: METHODS AND PROBLEMS

The most direct measure of socioeconomic impact is employment. The number of jobs robots could potentially take over does not, of course, translate directly into unemployment. In a growing economy displaced employees can simply be shifted to other jobs. The social problem is one of transitional assistance and retraining. Even if the economy does not grow, the natural rate of attrition of several per cent per year allows leeway for a moderate

rate of robot introduction without any displacement. In this case the problem for society is to provide sufficiently many entry-level jobs in other sectors of the economy.

TABLE 7 Maximum Market Growth Rate (% p.a.)

Discount rate δ	Experience parameter a	Initial capital payback period (years) 2	3	5	6.67	10
0.08	0.2	130.5	77.7	37.3	21.8	6.2
	0.3	81.6	48.6	23.3	13.6	3.9
0.12	0.2	118.1	65.3	24.8	9.3	*N*
	0.3	73.8	40.8	15.5	5.8	*N*

N: No growth is possible

Methodological issues arise primarily in connection with the question: How will the introduction of robots affect productivity of labour? of capital? How will it affect overall economic growth? There is no doubt that robots will increase the efficiency of the manufacturing sector, but it is not immediately clear whether reduced labour and capital costs in manufacturing will then be translated into additional employment generating demand for services. Which sectors (if any) will grow? Of course, full employment can be maintained artificially, but can we be sure that real wages would grow?

We are attempting to analyse these issues with help from a sequence of economic models of increasing complexity. The first step in the analysis is an examination of the productivity issue in the narrow context. The survey of robot users and potential users, mentioned earlier, asked respondents to explain their motivation for introducing robots. The reasons given were as follows:

reduce labour costs;
improve product quality;
eliminate dangerous or unhealthy jobs;
increase output;
reduce materials cost (e.g. in spray painting);
increase flexibility (e.g. in connection with a mass production operation.

Only the first three motivations were considered primary (i.e. sufficient) by any of the respondents; the last three were considered to be secondary benefits. Despite the emphasis on labour costs—especially in the automobile industry, which currently has wage costs (direct and indirect) of $20 per man-hour—we found that the most significant impact of robotics is likely to be to increase capital (especially machine tool) utilization and thereby reduce *batch* manufacturing costs more nearly to *mass* production levels.

The cinquennial *American Machinist* inventory of machine tools gives data on the number of machine tools, by type, in each sector of the metal-working industries. The occupational employment data mentioned previously, from the Bureau of Labor Statistics, provides data on the number of *machine operators*, by type of machine and by sector. Combining the two we derived estimates of machine tool utilization in metal-working. Assuming each worker works an 8-

hour shift, we find the following (for manually controlled tools:)

metal-cutting tools	12%
metal-forming tools	15%
joining (e.g. welding) equipment	22%

The actual figures are even lower, since much machine time is actually unproductive due to set-up, gauging, loading/unloading, tool changing and equipment failure. The figures for NC machine tools are better, since multi-shift operation is more common. But dramatic improvements in capital utilization can be achieved by integrating NC machine tools served by robots, into "cells" and combining cells into computerized "flexible manufacturing system" (FCMS).

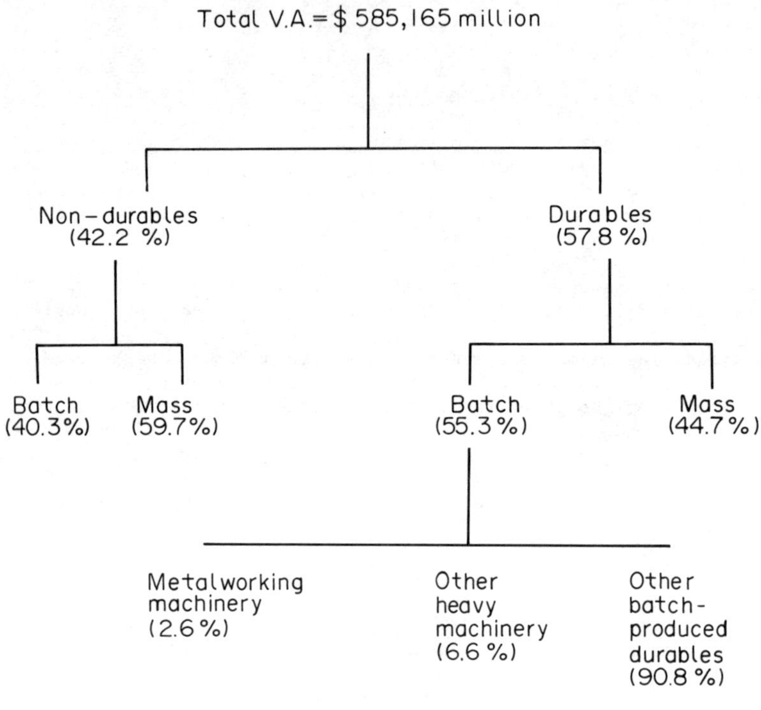

Figure 3. Distribution of manufacturing value-added ($US 1977).

Subdividing the manufacturing sector roughly into "batch" and "mass" production (Figure 3), it can be seen that durable goods, in particular, are predominantly manufactured in batches. We have estimated, on the basis of crude empirical comparisons between unit costs in batch and mass production, that extensive use of computers and robots in manufacturing can cut unit costs of batch-produced durable goods—including most capital goods—by 25% or so, in real terms! This would clearly have a beneficial "ripple effect" on the rest of the economy: potential long-run labour savings in manufacturing labour force and of the order of 5% of the national labour force. Assuming this occurs over two or three decades, the displacement trauma should be minimal.

Turning to the issue of economic growth, the simple Harrod-Domar model can be expressed as follows:

$$\dot{Y}/Y = S/R \tag{1}$$

Where Y is the aggregate output of the economy (GNP), S is the aggregate savings rate and R is the aggregate capital/output ratio. Assuming the savings rate is unaffected by the introduction of robots, some positive impact on growth would seem to follow from the increased capital utilization mentioned above. (Of course, machine tools are only a small part of the overall economy.) The fact that capital per unit of labour has been increasing tends to disguise the fact that the capital/output ratio has been decreasing in the US for a long time (e.g. from 4.58 in 1919 to 2.262 in 1960).[7] Increased use of robotics would merely accelerate this trend.

Given that the impact on economic growth would be positive, what can be said about real wages? Here, the relevant analysis was carried out long ago by Herbert Simon, in a study of the economic impact of information technology.[7] Using a simple static one-sector model Simon shows that under realistic assumptions, if the real rate of interest remains constant technical improvement must result in increasing real wages. Simon argues that the real rate of interest *cannot* increase as a consequence of technological changes of the type being considered; i.e. changes that reduce amount of capital needed per unit of output. In fact, the simple model shows that any technological improvement, regardless of whether it conserves capital or labour, will increase real wages.

We extended the one-sector model of Simon, to distinguish capital from consumption goods, and "dynamicized" it by introducing explicit population change, technological improvement (expressed in terms of a production function) and investment.

$$Y = p_c C + p_I I \tag{2}$$

Where C, I are outputs of consumption and capital sectors, respectively

$$C = F_k(K_k, L_k) \tag{3}$$

$$I = F_c(K_c, L_c) \tag{4}$$

$$L = L_c + L_k \tag{5}$$

$$I = I_c + I_k \tag{6}$$

Only the price ratio $p = p_c/p_I$ is significant. (In fact, we can arbitrarily set $p_1 = 1$, whence $p = p_c$.) Equating the marginal productivity of labour and capital in each sector one obtains

$$\frac{p_c}{p_1} = \frac{\partial F_k/\partial L_k}{\partial F_c/\partial L_c} \tag{7}$$

$$= \frac{\partial F_k/\partial K_k}{\partial F_c/\partial K_o} \tag{8}$$

Capital accumulates according to the rules

$$K_k(t) = (1-d_k)K_k(t-1) + I_k(t-1) \tag{9}$$

$$K_c(t) = (1-d_c)K_c(t-1) + I_c(t-1) \tag{10}$$

Finally we retain the constant savings propensity assumption: in the form:

$$I - sY \tag{11}$$

The labour force (or population) increases independently of the economic system, by the rule

$$L(t) - (1+g)L(t-1) \tag{12}$$

If we now assume Cobb-Douglas production function of the form

$$F_k = a_1(AK_k)^{a_2} L_k^{\;1-a_2} \tag{13}$$

$$F_c = b_1 K_c^{\;b_2} L_c^{\;1-b_2} \tag{14}$$

the model can be reduced to two independent equations by algebraic manipulators. We programmed the model for a PDP-11 computer and generated a number of sample "runs". Not surprisingly, this more complex model confirmed the qualitative conclusions of the simpler model.

More complex models are obviously needed to deal with issues of structural change *within* the economy. From a public policy point of view it is of great importance to anticipate which sectors will grow (and which will decline) in order to devise intelligent strategies for education, retraining transitional assistance and public sector investment. Descriptions of several multi-sector models to deal with some of these questions are given in Appendix A.

TRENDS IN THIS KIND OF EXERCISE

I can offer no more than an impression of trends in the study of "futures". My first impression is that two decades ago (or even less) such an analysis would have been much less quantitative and much more speculative. For instance, there would have been more reliance on historical analogies (e.g. the "industrial revolution") and little use of economic models.

One reason that quantitative models are used increasingly is that data are more readily available. The quality and quantity of data available for economic analysis in the US have increased very dramatically since World War II, and especially since 1969. (In the rest of the world the improvement is even more dramatic.)

Another reason for the shift is that economics has progressed and economists have relatively higher prestige than was the case 20 years ago. The awarding of Nobel Prizes in economics has greatly increased the prestige of the profession.

A third reason for the shift is that computers have improved so much—and have become so widely available—that large-scale empirical analysis is relatively cheap and easy. Data-intensive jobs that would have required many man-years 20 years ago can now be undertaken almost casually.

The second trend I think I see is a reaction against building very large national or global models. The increased availability of both data and computers made such models possible, but I think we have mainly learned to distrust them—at least as forecasting tools. The problem, I think, is that we only know how to quantify a few of the significant phenomena involved in social change. A large model necessarily depends entirely on these few quantifiable factors. Yet large size generates a misleading impression of "completeness", while disguising the details of its internal workings from all except the builders. The audience, having been too impressed at first, is now in a very sceptical mood.

Nevertheless, I suppose large-scale models will eventually regain a place in the arsenal of forecasting methodologies.

NOTES

1. Griliches, Zvi. Price indexes and quality change. In: *Studies in New Methods of Management*, Harvard University Press, Cambridge, Mass., 1981.
2. Sahal, D. A theory of measurement of technological change. *Int. J. Systems Sci.*, 8, 1977.
3. Saaty, T.L. *The analytic hierarchy process*, McGraw-Hill, New York, 1980.
4. Newell, Allen. Heuristic programming: ill-structured problems. In: Aronofsky, J.S. (ed.), *Progress in Operations Research*, vol. III. Wiley, New York, 1969.
5. The closest thing to such a measure, so far, is probably the chess-playing skill rating system. Based on this system a master has a rating of 2200, a "grand master" has a rating in the range of 2500 and a world champion might be rated at 2600-2700. The most advanced chess programs are currently rated at about 2200 (on the master borderline).
6. For a good summary of the method, see Cunningham, James A. Using the learning curves as a management tool. *IEEE Spectrum*, June 1980, pp. 45-8.
7. See Simon, H. *The New Science of Management Decision*, Prentice-Hall, Englewood Cliffs, NJ, 1977.

APPENDIX

A. Maindirotta

A.1.0 GROWTH MODELS FOR PRODUCTIVITY IMPACT ANALYSIS

The two important features that it is desirable that the growth model should capture and evaluate are as follows:

(1) possibilities for improvement in the economy in terms of real consumption and real growth resulting from improved productivity of capital in the manufacturing sector,

(2) structural shifts in the economy induced through relative price changes—in consumer demand patterns, in investment allocation across sectors and in production technologies employed (in terms of the relative proportions of capital and labour).

The preliminary cause, whose effects it is desired to trace, is improving capital productivity in the manufacturing sector of the economy. A two-sector model that divides the economy into manufacturing and non-manufacturing sectors naturally suggests itself. Ultimately, however, interest centres around the output of investment and consumption goods. The division of the economy into manufacturing and non-manufacturing sectors does not correspond to the division of the final output into investment and consumption goods. This is partly due to inter-industry transactions and partly because both investment and the consumption basket include manufactured and non-manufactured goods.

There are also other problems with this division. Within the non-manufacturing sector the diversity of outputs and production technologies is so large that it appears very unrealistic to describe this sector's output through an aggregate production function. It may be desirable to disaggregate this sector suitably along some dimension of production technology. Within the manufacturing sector, production technology is more homogeneous and an aggregate production function is not so objectionable. However, the introduction of CAM and robots can be expected to have differential impacts in the several manufacturing subsectors. In particular it is conjectured that the impact, in terms of improved productivity of capital, would be substantially different between those subsectors that produce capital goods and those that produce consumer goods. Since the changing relative cost of production of investment and consumption goods significantly influences the path of the economy, it appears desirable to disaggregate the manufacturing sector along the dimension of potential for productivity gains.

The problems created by the non-coincidence of the investment/consumption goods division and the manufacturing/non-manufacturing sector division, and the further disaggregation of the manufacturing and non-manufacturing sectors, can be handled by various means. These include introducing input-output coefficients and assuming that the homogeneous investment good is produced effortlessly by combining manufacturing and non-manufacturing outputs in definite proportions. Consumption can be similarly handled, or more realistically treated as a vector quantity.

Such devices are, however, not always convenient. For example, in one of the models suggested hereafter, introduction of a multidimensional utility function (the arguments being the consumption vectors) having any other than an additive form would destroy the linear structure of the model, making it much more difficult to solve. In a second model, introduction of inter-industry transactions would complicate computation of the marginal return on investment across sectors—equalization of which is the investment allocation rule.

Other specification issues to be resolved include:

(1) the treatment of capital as

   (a) homogeneous (i.e. the capital stock used in any one sector being indistinguishable from that used in any other sector—which effectively means that the economy produces only one investment good) or non-homogeneous,

   (b) and if homogeneous, as mobile (i.e. capital can be freely transferred across sectors—even after installation) or immobile (i.e. only fresh investment can be freely allocated across sectors);

(2) the treatment of the technical progress resulting from the introduction of CAM and robots as embodied (i.e. affecting only new investment) or disembodied—which amounts to treating capital as vintage or non-vintage respectively;

(3) the possibilities for factor substitution, i.e. whether the model should be a putty-putty model or a putty-clay model or a clay-clay model;

(4) the treatment of depreciation as occurring by evaporation or by sudden death (one-horse shay). Depreciation by evaporation means "that of any given stock of machines, no matter how old or new they are, a certain percentage collapses or disappears each year". Depreciation by sudden death means that the machines have a fixed lifetime through which their productivity remains constant.

A putty-clay model with non-homogeneous, immobile capital and some degree of both embodied and disembodied technical progress appears the closest description of reality. (As pointed out earlier, while the introduction of CAM and robots is likely to make some contribution to the productivity of existing installations, the greatest impact is expected in essentially new facilities designed from the ground up.) However, such a formulation may prove quite intractable and one or the other of the above conditions may need to be modified. Three alternative formulations are given in the following subsections. Given below are some comments on these formulations.

The first model, a multi-sector programming model (probably 6–10 sectors at most), assess the additional benefit to society—in terms of an aggregate utility function whose argument is the time stream of consumption—that would be possible with the introduction of CAM and robots in the manufacturing sectors. It is assumed that the allocation of the final output in each period between consumption and investment across sectors can be controlled, and that the economy is adequately described by an input-output structure with fixed coefficient, primary input (capital and labour), production functions for each sector. The potential for productivity gains from the introduction of CAM and robots is also assumed to be differential across sectors. Capital is treated as vintage and immobile, and the investment good proportionately composed of the output of the several sectors. If an additive utility function is assumed, the model has a simple linear structure that makes solution easy.

The model can address employment concerns in one of three ways:

(1) It can be assumed that labour availability is never a constraint, and the maintenance of employment levels is of secondary concern (as does the formulation of Section A.2.0). Given output-labour coefficients, the implications for employment over time for the two optimal solutions (with and without capital augmenting technical progress in the manufacturing sectors) can be examined.

(2) An autonomous growth rate of population can be specified, and the maintenance of given levels of employment, in each period, imposed as constraints. (It would be interesting to compare the constrained and unconstrained solutions to see the trade-offs, if any, that are involved.)

(3) The level of employment can be introduced as an argument in the utility function.

To assess the structural shifts induced in the economy as a result of productivity gains in the manufacturing sector requires a descriptive model that endogenizes the structural variables. The second model has only two sectors: capital goods and consumption goods producing sectors. It projects the path of the economy as a whole to capture price-induced shifts in production technology and the allocation of investment across sectors, that would result from the introduction of CAM and robots primarily in the capital goods sector. Of particular interest, however, is the induced shift in production technology in the consumption goods sector.

The model uses the investment allocation rule of equalizing the marginal return on capital in the two sectors. This is made simple through using linear homogeneous production functions in both sectors. These functions have the convenient Eulerian property that returns to factors of production on the basis of their marginal productivity exhaust the total product. The model has a "putty-putty" structure, treats capital as immobile, and technical progress as disembodied and capital-augmenting.

One of the impediments to growth in the US economy in recent decades has been a low savings rate as compared with other industrial economies. An interesting question arises: how much improvement in the growth rate could result from improving productivity in the capital goods sector if the savings rate remains constant? The second model particularly illuminates this issue.

As the model is presently formulated (Section A.3.0) it is a full employment model. Production decisions are taken so as to optimally utilize the complete stock of capital and the total supply of labour—the rate of population growth being assumed to be autonomous. By imposing a constraint on the minimum real wages, the model can be modified to reflect the impacts of the introduction of CAM and robots on the overall levels of employment over time.

Admittedly, the representation of the economy as composed of only two aggregate sectors—the capital goods and consumption goods producing sectors—with no intermediate outputs, and with only the capital goods producing sector benefiting from the introduction of CAM and robots, is less realistic than the representation in the first model described above. The problem is that incorporating inter-industry transactions would destroy the simple structure of the second model, making it difficult to evaluate the marginal return on capital. Some other investment allocation rule would then be needed. Model formulation in such a case would then be quite difficult, as demonstrated by the fact that investment behaviour remains the weakest link in macrodynamics. (As long as inter-sector transactions are excluded, the model can easily accommodate further disaggregation of the consumption goods sector—all that would need to be added is consumption functions for the outputs of the individual subsectors.) Given the interesting issues this model addresses, the oversimplifications of reality may be worth making.

The third model is one whose conceptual scheme, only, is presented. The appropriate mathematical formulation has not yet been completed and it is not yet certain that a tractable formulation will be possible. Conceptually it attempts to overcome some of the drawbacks of the second model—namely it replaces disembodied by embodied technical progress and the "putty-putty" structure by a "putty-clay" structure. That is, for plants installed in period t, factor substitution is possible at the planning stage—period t—and factor combinations are chosen to optimize production in the two sectors given the prices of products and the wage rate of labour. After installation the production technology of a plant is assumed to be frozen for all time, and the plant remains operative until it either exceeds its assigned lifetime or becomes unprofitable at the current prices and wages.

The difficulty with this model arises from the fact that the labour supply is dependent on the prices of the products, and the wage rate in an essentially discontinuous fashion (due to the labour released by unprofitable plants). The prices of the products and the wage are in turn dependent upon the labour supply through the optimization involved in the planning of output from new plants. The solution of the model for even a single period, within this general equilibrium framework, would require iterative procedures and would be quite tedious. Multiperiod solutions become correspondingly more difficult. However, there appears to be a possibility of making solution easier by making producer behaviour dependent upon previous period prices and wages—an assumption that is quite plausible. This enquiry is left as one of the proposed research tasks.

A.2.0 MODEL 1

A.2.1 Assumptions

(1) An m-sector economy with fixed coefficient production functions in each sector—i.e. output in each sector is a fixed multiple of capital stock.

(2) The introduction of CAM and robots manifests itself as autonomous/ embodied technical progress resulting in changing output-capital co-efficients over time (for fresh investment—i.e. new facilities only—a result of the "embodied" technical progress assumption). The impact is differential across sectors.

(3) The capital stock is assumed to have a fixed, exogenously specified lifetime in each sector.

(4) Inter-industry transactions are suitably modelled through a fixed co-efficient input-output structure. These transactions include maintenance requirements to keep old capital productive until its lifetime is up. Consequently, the final output of each sector is "net" output, available for fresh investment and consumption.

(5) There is one homogeneous investment good that is produced by effortlessly combining the final output of the m sectors in definite fixed proportions. However, investment is not mobile—i.e. only fresh investment can be freely allocated across sectors.

(6) There is a 1-year investment to output lag—i.e. investment composed of the output of period "t" is productive only from period $t+1$" onwards.

(7) Consumption is a vector quantity. The utility function is monotonic in the consumption of all commodities.

(8) The model is a planning model; the policy dimension being the allocation of the final output in each period between consumption and investment in the m sectors; the objective being maximization of utility from the time stream of consumption.

(9) Labour availability is never a constraint, so that it can be entirely ignored.

A.2.2 Definition of Terms

(Notation: An upper case letter enclosed in [ ] is a matrix. An upper case letter underlined is a column vector. The addition of a prime transposes the given column vector into a row vector. An expression enclosed in ( ) is the typical element of the associated vector or matrix.)

$\underline{O}_t = (o_t^j)$ $j = 1$ to m; where o_t^j is the output of the jth sector in period t

$\underline{F}_t = (f_t^j)$ $j = 1$ to m; where f_t^j is the final output of the jth sector in period t

$A = (a_{jk})$ $j,k = 1$ to m; where a_{jk} is the amount of the jth good required as input for production of one unit of the kth good

$\underline{C}_t = (m_t^j)$ $j = 1$ to m; where c_t^j is the amount allocated for consumption, in period t, out of the final output f_t^j

$\underline{M}_t = (m_t^j)$ $j = 1$ to m; where m_t^j is the amount allocated for investment, in period t, out of final output f_t^j

TI_t is the total investment good produced in period t

$\underline{I}_t = (i_t^j)$ $j = 1$ to m; where i_t^j is the amount of investment good TI_t

allocated to sector j in period t

$B_t^j = (b_s^j)$ $s = t - 1$ to $t - 1(j)$; where b_s^j is the output-capital coefficient in the jth sector for investment made in period s and $1(j)$ is the lifetime of capital in the jth sector

$\underline{K}_t^j = (i_s^j)$ $s = t - 1$ to $t - 1(j)$; where i_s^j is the investment made in sector j in period s

d^j is the proportionality factor that relates total investment TI_t to allocations m_t^j.

A.2.3 Mathematical Statement of the Model

Maximize $U(\underline{C}_1, C_2, \ldots \underline{C}_\infty)$
over
$(\underline{I}_1, \underline{I}_2, \ldots \underline{I}_\infty)$

subject to

(1) Input-output consistency conditions

$$\underline{O}_t = \underline{F}_t + [A] * \underline{O}_t$$

(2) Output-capital relationships

$$o_t^j < \underline{B}'^j_t * \underline{K}_t^j \text{ for } j = 1 \text{ to } m$$

The initial distribution of capital stock may be such that, in the first few periods, a vector of full capacity outputs may not satisfy the input-output conditions. As such the above relationships have to be expressed as inequalities.

(3) Allocation of final output for investment and consumption

$$\underline{M}_t + \underline{C}_t = \underline{F}_t.$$

(4) Proportionate composition of investment condition

$$\underline{TI}_t = d^j * m_t^j \text{ for } j = 1 \text{ to } m$$

The above two conditions can be written as strict equalities since any surplus can always be taken up by consumption to increase the value of the objective function (which is assumed to be monotonic in the consumption of all commodities).

(5) Allocation of total investment made in period t across sectors

$\underline{TI}_t = \underline{S}' * \underline{I}_t$ where $\underline{S}'$ is the $(l * m)$ sum vector $(l, l \ldots 1)$.

(6) Non-negativity conditions

$$\underline{I}_t \geq 0$$

In the absence of this condition, the program may make negative investments in some sectors, effectively treating capital as mobile. Since capital is to be treated as immobile this condition has to be imposed.

Given $\underline{B}^j$ and $\underline{K}_0^j$ for $j - 1$ to m as initial conditions

A.2.4 Comments

The model is an optimization problem best handled as a programming problem—which brings one up against the standard "horizon" problem. Essentially what we have is an infinite horizon. However, since consumption in distant time periods is expected to contribute very little to overall utility, a solution to a finite horizon reformulation of the problem, with specified terminal conditions, should be a close approximation to the original problem. Sensitivity analysis to the length of the horizon and the terminal conditions can ensure a reasonable result.

The size of the problem than can be handled, in terms of the degree of disaggregation—i.e. the number of sectors—and the length of the horizon, depends upon the formulation of the utility function. An additive utility function will result in a non-linear program with a non-linear objective function and linear constraints—in which case the size of the problem than can be reasonably handled is much smaller.

Model specification and model solution are thus related issues involving trade-offs. Exploring these trade-offs will constitute a major research task.

A.3.0 MODEL 2

A.3.1 Assumptions

(1) A two sector economy (capital goods and consumption goods producing sectors) with two-factor (labour and capital) linear, homogeneous production functions, permitting factor substitution in both sectors.

(2) The introduction of CAM and robots manifests itself as autonomous/disembodied capital augmenting technical progress—primarily in the capital goods production sectors.

(3) The capital good is homogeneous, though immobile once installed.

(4) Gross investment (in real terms) in period t is essentially the output (in real terms) of the capital goods sector in period t.

(5) There is a 1-year investment to output lag—i.e. investment made in period "t" is productive only from period "$t + 1$" onwards.

(6) Capital depreciation rates are fixed and exogenously specified for each of the two sectors.

(7) The economy has a constant savings propensity (in money terms).

(8) The model is a full employment model. There is no independent investment demand function. The money value of investment is just the money value of savings.

(9) Investment between sectors is so allocated that the marginal return per unit of physical capital is the same in both sectors.

(10) The labour force grows at an autonomous rate and labour is so allocated that its marginal productivity is the same in both sectors. This marginal product is also essentially its wage rate.

A.3.2 Definition of Terms

I_t is the output of investment goods in period t

C_t is the output of consumption goods in period t

K_t^k is the capital stock in the capital goods sector at the beginning of period t

K_t^c is the capital stock in the consumption goods sector at the beginning of period t

L_t^k is the labour employment in the capital goods sector in period t

L_t^c is the labour employment in the consumption goods sector in period t

L_t is the total labour supply in period t

p_t^I is the price of investment goods in period t

p_t^c is the price of consumption goods in period t

d^k is the depreciation rate in the capital goods sector

d^c is the depreciation rate in the consumption goods sector

F_k^k is the marginal physical product of capital in the capital goods sector

F_k^c is the marginal physical product of capital in the consumption goods sector

F_I^k is the marginal physical product of labour in the capital goods sector

F_I^c is the marginal physical product of labour in the consumption goods sector

Y_t is the total money value of output in period t

s is the constant savings propensity (in money terms)

A.3.3 Mathematical Statement of the Model

Production function relationships

Capital goods sector

$$I_t = F^k(A(t) * K_t^k, L_t^k) \tag{1}$$

where $A(t)$ is the -shaped productivity gain relationship

Consumption goods sector

$$C_t = F^c(K_t^c, L_t^c) \tag{2}$$

Wage rate (marginal product of labour) identity relationship

$$[p_t^c * \frac{\partial F^c}{\partial L_c}]_t = [p_t^I * \frac{\partial F^k}{\partial L_k}]_t \tag{3}$$

Labour accounting identity

$$L_t^c + L_t^k = L_t \tag{4}$$

Labour growth equation

$$L_{t+1} = (1 \quad G) * L_t \tag{5}$$

where G is the percentage annual growth rate of the labour force

Capital accumulation equations

Capital goods sector

$$K_t^k - K_{t-1}^k + I_{t-1}^k - d^k * K_{t-1}^k \tag{6}$$

Consumption goods sector

$$K_t^c = K_{t-1}^c + I_{t-1}^c - d^c * K_{t-1}^c \tag{7}$$

Investment allocation identity

$$I_{t-1} = I_{t-1}^k + I_{t-1}^c \tag{8}$$

Investment allocation rules

$$[p_t^c * F_k^c]_t = [p_t^I * F_k^k]_t \tag{9}$$

i.e. the marginal return condition

Total income

$$Y_t = p_t^c * C_t + p_t^I * I_t \tag{10}$$

Constant savings rule

$$p_t^I * I_t = s * y_t \tag{11}$$

A.3.4 Comments

The unknowns in each time period are I_t, C_t, K_t^k, K_t^c, $(L_t^k$, L_t^c, L_{t+1}^k, I_{t-1}^k, I_{t-1}^c, Y_t and relative price (p_t^c/p_t^I)—eleven in all. There are also eleven relations. Hence, given the prior period values, all the variables in a particular period are determined. Thus, given the initial conditions, the complete time path of the economy is determined. The mode can hence be run as a simulation model.

A.4.0 MODEL 3

A.4.1 Assumptions

(1) A two-sector economy-capital goods/consumption goods.

(2) The capital good is homogeneous, though immobile once installed.

(3) Output in both sectors in period 5 comes from two sources:

(a) plants installed in period t using as investment the output of capital goods in period $t-1$,

(b) plants installed in prior periods.

(4) Plants installed in period t are installed choosing the optimal capital intensity, given a linear homogeneous "planning" production function that allows for substitution of capital and labour. As the model is presently formulated these are current period prices and wages.

(5) Output from plants installed in prior periods is predetermined on the basis of decisions taken at the time of installation—i.e. their capacity and labour requirements (in physical terms) is fixed. However, only those plants which have not exceeded their given lifetime *and* are profitable at the given prices and wage rate, produce output in period t.

(6) The introduction of CAM and robots results in the planning production function in the capital goods sector shifting overtime. Technical progress is autonomous and embodied.

(7) Suitable consumption and investment functions determine the demand for consumption and capital goods in period t, in response to prices and income of period t.

(8) Population grows at an autonomous rate.

A.5.0 MODEL PARAMETERIZATION

The functions and parameters that need to be determined in order to complete model specification are essentially:

(1) For Model 1:
(a) the input-output and the output-capital coefficients,
(b) capital lifetimes in the several sectors,
(c) the utility function.

(2) For Model 2:
(a) the aggregate production functions in the two sectors,
(b) the capital depreciation rates in the two sectors,
(c) the aggregate savings propensity of the economy.

(3) For Model 3:
(a) output-capital coefficients for plants of different vintages,
(b) aggregate "planning" production functions,
(c) investment and consumption functions.

It must be emphasized that, given the nature of the enquiry and the abstractions already involved in model formulation, it would be worthless to get any better than gross estimates for all of the above. Wherever there is considerable ambiguity about the true value of a parameter, sensitivity analysis could be resorted to, to get an idea of its importance.*

Two sets of values will be needed for all of the parameters. The first are values expected to prevail in the absence of any progress towards the introduction of CAM and robots in manufacturing processes. The second are the values expected under such progress. For simplicity, the assumption of no change is made for the former and existing values are taken to be the ones that will continue to prevail. Since the introduction of CAM and robots, characterized as capital augmenting technical progress, is assumed to directly influence only capital productivity in the manufacturing sector's specification of the -shaped productivity curve discussed earlier along with the existing values for all the other parameters, should completely describe the second context. Technological forecasting procedures will be needed to specify the curve. Some research will be needed to determine the most appropriate procedures.

The conditions under which a utility function can be expressed in a particular functional form, and the corresponding procedures for its measurement, are well known. Determination of the utility function would involve selecting one of these as representative of the "policy-maker's" preferences.

The strategy suggested to determine the values of the parameters, such as the input-output and capital-output coefficients, is to derive them from the parameters of earlier, more disaggregated macroeconomic studies. These sources could be supplemented by data published by the Bureau of the Census, US Dept. of Commerce, in the "United States Census of Manufacturers" and the Bureau of Labor Statistics, US Dept. of Labor.

*In any case, some of the constructs used in the models have doubtful empirical status. The aggregate production function is well known to be one. Either one believes that such a theoretical construct has something to contribute to our understanding of the economy or one does not. This inquiry is clearly based on the former viewpoint.

FHN - F*

Scientific and technological forecasting and higher education

O. M. Bielotserkovski

The present phase in the development of human society is justly called the era of the scientific/technological revolution. Its characteristic features are an impetuous flight of scientific thought, rapid technical progress and basic qualitative changes in all spheres of activity of man, from elementary technological operations to the management of large industrial complexes.

In recent years new experimental discoveries and theoretical concepts promoted the development of fundamental and applied branches of science. Epoch-making discoveries in science have been made in microphysics, nuclear energetics, molecular biology, cybernetics and in the exploration of space. At both extremes of scale the exploration of space has been typical of the conduct of modern science.

The exploration of the cosmos has been a significant landmark in the development of natural sciences. Nowadays, astronautics has become a major branch of science and technology, and as a result our knowledge of the earth, the moon and the other planets of our solar system is unmeasurably greater than previously. Sometimes a 5-year period proved sufficient (as was the case in the exploration of Mars) to refute numerous hypotheses and guesses—which had persisted for a long time—and completely adopt new scientific concepts.

Progress in scientific thought has been the basis for constructing fundamentally new techniques. Modern complex machinery, supersonic passenger airliners, gigantic turbines, atomic power stations and atomic powered icebreakers—all these have been created within the walls of research institutes and the design offices. One of the striking examples of the influence of scientific thought on technical progress is the field electronic computers. Being a logical outcome of the advances in such fields of science as mathematical logic, theory of information, and electronics, the computer is now an important factor in technical progress. If computers did not exist we could not solve such technical problems as splitting an atomic nucleus or launching earth satellites. When implemented in the production process they comprise a qualitatively new stage in the automation of the sphere of production management.

The scientific/technological revolution is, indeed, a global process taking place in all countries of the world. But it acquires some special features under the conditions of socialism. Socialism is a social system under which science is given a really humanistic orientation and its results are at the service of humanity. Within the structure of science is developing on prescribed lines. This systematic development of science and the application of its results for the benefit of society is a state policy of the USSR, and is reflected in article 26 of the Constitution of the USSR which says: "In accordance with the needs of society, the state provides for planned development of science and the training of scientists and organizes introducing the results of research into the economy and other spheres of life."

The scientific/technological revolution has a profound effect on all levels of general and special education. The USSR and its higher schools are now confronted with new, more complex problems. A constant expansion of the front of scientific/technological developments, and the necessity of their practical realization in industry, require an influx of young specialists with thorough theoretical and professional backgrounds who are capable of creatively managing modern production and seeking optimal solutions in practical work and in sciente through the treatment of huge quantities of information.

A cardinal problem of special higher and secondary education now appears to be that of the quality of the training of specialists. The very notion of the "quality of training" results from the importance attached to the professional and sociopolitical activity of the specialists in solving a broad spectrum of the problems of modern socialism. These problems are the unprecedented scale and rate of development of the socialist national economy, and its more complicated structure and interbranch contacts; a wide implementation of the scientific results in the practice of applied programme; emphasis on a more complete utilization of the intensive factors of economic growth, and so on.

The increased requirements for the preparation of specialists are also due to the accelerated development of social processes, the greater influence of subjective factors in the realization of the laws of social development, the rapid proliferation of the scientific principles of management and control in all spheres of social life; new possibilities for improving social relations in order to achieve complete social equality, and comprehensive development of the personality of man. These and other trends taken together form an initial set of criteria for the quality of the training of specialists, which predetermines *principally new* directions for improving special higher and secondary education.

The peculiarity and, to a certain degree, the urgency of this problem lies in the fact that it is associated with the system of education oriented to a wide strata of population.

In the context of the scientific/technological revolution new scientific directions are established and develop very rapidly. Thus, we feel the need for specialists for these new directions, and one of the major requirements is *professional flexibility* on the part of the specialists. Generally speaking, there is a certain contradiction between the increasing real and potential requirements of the specialists, and the somewhat conservative nature of the educational system, which tends to be less flexible than rapidly developing science and production. To adequately assess this process of development and propose appropriate directions for the preparation of the specialists we believe that *a fundamental reorientation of higher education, and its transition from an extensively informative to an intensively fundamental system of instruction are necessary.* Within this framework, a new approach to the solution of the problem concerning the profile of specialists expressed by the formula "narrow specialization—on broad basis" is formed.

The whole system of education must be aimed at the fundamental and methodological training of specialists in a definite area of knowledge. On this basis, one must develop the creative talents and interests of the students, and the ability to acquire independently new scientific/technological and sociopolitical information. The ability of a specialist to see the perspectives of the development of his branch of science and the country's economy as a whole, and to correctly solve the problems of the scientific organization of labour and production management, is of major importance.

However, not only industry but also science (as we have mentioned, a more dynamic system than education) require that the structure of the preparation of specialists be sufficiently flexible in order to educate the specialists with the necessary profile in as short a time as possible.

For this reason specialized departments and higher schools of a new type, i.e. *technical universities* to train specialists in various fields of modern physics and technology, were developed in the USSR. At present, a similar situation can be seen in the training of specialists in medical physics, quantum electronics, perspective energetics, computer design, and economists of a broad profile, applied mathematicians, physicists in technology, physicists-designers and representatives of some other specialities are seen to be of top priority for the country.

Specialists of very diverse profiles are now confronted with *complex* scientific, engineering and technological problems that are closely interrelated.

For instance, space exploration, the realm of science which is closest to my own scientific interests, set forth before engineers and scientists/aerodynamicists a whole series of quite new problems related to the determination of the aerodynamic characterists of rockets and re-entry vehicles. Here we have a deep interpenetration of purely scientific problems (usually related to the formulation and solution of complicated non-linear problems of physical gas dynamics), applied problems (the development of efficient solution methods and their realization), engineering problems (the determination of most successful solutions from the viewpoint of a whole system using theoretical and experimental results) and purely production-related problems dealing with the realization of a project.

Recent years represent a stage of intensive development of diverse scientific and engineering methods of investigation; and of crucial importance in theoretical developments was a broad practical implementation of investigations involving high-speed computers.

Now it is necessary to train analysts and practical scientific workers so that they have a profound theoretical and engineering knowledge and will be able to solve complex scientific/technological problems arising at the junction of various fields of knowledge.

Finally—and this aspect is of special significance—at the modern stage of the scientific/technological revolution, practical work requires advanced forms of cooperation between the higher schools and science and production. This requirement is one of the main principles in the development of higher schools in the USSR.

For the effective solution of the above problem, constant research and development of new forms of higher education are necessary. Here it is appropriate to make a small excursion into history.

Different social epochs in the development of society set down different requirements for the form and content of education. More than 200 years ago, our famous compatriot M.V. Lomonosov, founded the first Russian University, on the threshold of the rapid development of Russia. The French Revolution promoted the creation of a unified state system of education and the organization of such outstanding higher schools as the Ecole Normale and Ecole Polytechnique (both founded in 1794). The Industrial Revolution in Russia in the second half of the eighteenth century led to the creation of the first higher engineering school upon which was modelled the Mining College (now the Leningrad Mining Institute) founded in 1773.

During the years of Soviet power, more than 40 million skilled specialists have graduated from the higher and secondary schools of the country. There are now 883 higher and 4383 secondary schools with a total of 9.85 million students.[1] In the USSR, there are two main lines in the training of specialists of high qualification: universities and technical schools. Broad training in general scientific subjects is characteristic of the first line, but the graduates, as a rule, are not completely prepared for work in industrial laboratories and research institutes. As for the technical schools, they do not provide a sufficiently broad education.

One now observes a certain merging of university and technical education. There have appeared a number of higher schools/technical universities—which combine the elements (and advantages) of both universities and technical schools. But now the tendency in most technical institutes is to combine in their curriculum the fundamentals of a university education with comprehensive engineering training in ultramodern branches of technology. The basic principle of these institutes is to train specialists in cooperation with research institutes, design offices and industrial enterprises, i.e. to organize education-science/production corporations. The Moscow Physics Technical Institute belongs to the same category of higher schools as the Moscow Electronic Machine-Building Institute, the Leningrad Lensovet Institute of Technology, Novosibirsk State University, and so on. It is the technical university that is, in our opinion, one of the forms of higher education that is best able to solve the problems posed by the modern scientific/technological revolution.

If the basis for the training of specialists were to be briefly formulated, then it must be as follows:[2]

(1) fundamental nature of education, introduction of research activity to the education process;

(2) direction in the training of specialists;

(3) differentiating system of education (the training of specialists depending on the requirements and problems of a particular profile);

(4) the retraining of specialists.

Let us consider the above points separately.

FUNDAMENTAL NATURE OF EDUCATION AND RESEARCH WORK OF STUDENTS

While training specialists with high qualifications (engineers/analysts and practical scientists) in rapidly developing branches of science and technology, general education must be based on fundamental subjects. Mathematics, general and theoretical physics, social sciences and a foreign language are to be studied according to the university curriculum, so that the future specialists may actively master these subjects. Additionally, much attention must also be paid to the study of applied practical aspects of the programme (laboratory work, independent work, written reports, etc.).

The main task of this general education cycle is to provide for a profound knowledge of the fundamental subjects, in order to develop the students' creative abilities. In addition, the comprehensiveness of the program must not prevail over the ideological and creative aspect of education (herein we see the difference between instruction and education). Apparently, it is the fundamental nature of a general education cycle that lends harmony and depth to the education process.

The general education cycle is followed by a specialized one. It starts with lectures and laboratory work of a wide profile. Thus, a definite mental horizon in special education is attained. For instance, it is important for future engineers to read selected chapters of theoretical physics, to master computational methods already a part of the training process, and so on.

The third cycle concerns *investigation*. Its purpose is to prepare for a concrete speciality. This cycle should be carried out in well-equipped laboratories or research institutes (*base institutes*, as we call them). It is important that instruction at this stage be conducted by specialists actively engaged in a particular field of science and technology. A necessary condition in the training of a future engineer/analyst is to involve him in active independent research or production work within the training process. Thus, in the atmosphere of a creative scientific community, a future specialist acquires the qualities of an analyst which will be so useful to him in his work.

Note that the division of the process into three independent cycles (general education, special studies and investigation) allows free management of the flexible organization structure. This guarantees flexibility and direction in the training of specialists and is also a premise for a differentiating system of education.

The fundamental general education promotes the formation of a basis for students' specialization, practically, in all branches of technical-physical knowledge, including those which are completely new.

The thesis of the fundamental nature of education is now almost universally accepted. However, its actual realization encounters many difficulties. Firstly, modernization of the traditional instruction programme is required. Secondly, efficient instructors and well-educated and trained students are also necessary. (Many years' experience at the MPhTI demonstrated, in particular, that instruction in fundamental subjects should ideally be entrusted to teachers with a university education. The engagement of such instructors for general theory departments at technical institutes is a task with top priority. Thirdly (this is apparently one of the major aspects), fundamental general education programmes must be closely and naturally correlated to special subjects.

DIRECTION IN THE TRAINING OF SPECIALISTS

The higher school must be fully aware when and how many specialists of various profiles are to be trained and where they are to work. The activity of the higher school as a whole will be more effective and creative if the answers to these questions are clearly defined. One ought not to ignore the psychological orientation of the students. We believe that the system of base institutes should be introduced into higher schools on a wider scale: first, in the period of diploma work and, later (as the higher school and the base institute become more closely interrelated) during the process of instruction. Frankly speaking, the higher school has to assume extra obligations, e.g. the analysis of the need for specialists. However, for major higher schools (e.g. those that function as technical universities) it is quite within reach. In particular, we at the MPhTI attach great (if not crucial) importance to the problems of "demand and distribution" of the graduates and orient the senior students accordingly.

In recent years a number of new departments have been organized at the MPhTI without increasing the total contingent of the students: the Department of Physical and Quantum Electronics (1964), the Department of Aeromechanics and Flying Vehicles (1965), the Department of Management and Applied Mathematics (1969) (the first in the country), the Department of Physics and Energetics Problems (1976) and, finally, the Department of Physico-chemical Biology organized this year.

Thus, a combination of the fundamental nature of general technical education and orientation of the training of specialists (which also applies, by the way, to universities) is a very important trend in modern technical higher education. It is just this concept that defines the essence and form of higher education expressed in "oriented profile—on broad basis".

At present, we are confronted with an acute problem connected with an omnipresent gap between higher school curricula (specifying the knowledge of graduates) and the level of science and technology. It is sometimes set forth (or suggested) that, since the higher school curriculum cannot be changed overnight, this gap is practically inevitable in the conditions of rapid advances in scientific knowledge. We do not believe that such a view is well grounded. A system of education which organically combines instruction with research work within the framework of modern laboratories, leading research institutes and design offices is able to substantially reduce this gap or even completely eliminate it. The experience of some higher schools in the USSR which have close contacts with their base institutes shows that when special training of the students is undertaken directly at research institutes, then the reorientation of research in these institutes immediately changes types and substance of knowledge acquired by the students. As a result, a graduate is familiar with the latest achievements in the field of science he is to deal with. This system seems to automatically control the development of corresponding fields of science and technology (this is what is meant by "direction" in the training of specialists).

DIFFERENTIATING SYSTEM OF EDUCATION

The system of a technical university allows the rational training of specialists within a university or even a department in a thoroughly differentiating manner based upon the profile and needs in a particular category of specialists. By having a wide spectrum of special ties (ranging from purely academic to the exclusively engineering or even production-related) one can differentiate, with a view to increased effectiveness, the training of specialists by varying special and research cycles of education.

Concerning the training of research workers, it is reasonable to select them through competition, as is the case, for example, in enrolling in post-graduate courses. Obviously, it is not necessary that all the students should be engaged in research at the final stage of instruction; the institute should arrange that the quantity of analysts be 90-95% of the total number of graduates. A certain number of the graduates, for various reasons (illness, natural inclinations, disappointment in the specialty and, thus, lack of desire to work actively, etc.) turn out to be less well-prepared by the end of the course of studies. During the whole course of studies at the higher school, the students who are not willing (or are incapable) to continue studying should be systematically expelled or given a chance to begin production work directly by granting them an appropriate diploma. This "two-stage" system of education would allow us to attract, during the final research cycle of education, talented people from remote regions (as, for example, is the case with post-graduate courses), devise a more creative and demanding programme for this cycle, and so on. Obviously, this question deserves serious consideration.

The idea of selection by competition is the essence of the system at the Moscow Physico-technical Institute. For instance, in the third course the students take a final exam in general physics; in the second and fifth courses they undergo attestation. A similar selection is also made at Moscow State University and the Leningrad Lensovet Institute of Technology. It should be remarked that this staged competition selection of the students for a deeper specialization produces good results at some of the best colleges in Western Europe (Ecole Normale, France) and the USA (the Massachusetts Institute of Technology and the California Institute of Technology).

By all appearances, only the differentiating system of education results in a gradual increase of the quality in the training of specialists by naturally developing and extending the whole system of education. Again, this system is easier realized within a technical university.

RETRAINING THE SPECIALISTS

Since the quantity of engineers and scientists has considerably increased, a top-priority task now is the *qualitative* improvement of the knowledge of the specialists engaged in science and technology. A research worker, especially one in the applied branches of science, must study and increase his knowledge throughout his life.

At present, research institutes and new branches of industry have enough specialists of high qualification, due to the many years' intensive training of young specialists. Moreover, it must be recognized that the development of practical investigations lags behind the development of science. A high percentage of scientific results remain, for a long time, the exclusive property of research institutes. Workers in the applied branches of science fail to follow the rapid development of scientific investigations and rapid change of direction in fundamental science. Therefore, the knowledge acquired at a higher school becomes obsolete in no time. Because of the specific nature of work in industry and applied research institutes this phenomenon seems natural, but it is nevertheless rather alarming.

Evidently, along with the training of specialists in completely new branches of science and technology one should, on a wider scale, organize the retraining and probation of specialists with priority being given to those in the newest branches of industry and applied research institutes. It is important, in this sense, to organize the *systematic* retraining of the specialists with a higher education on burning questions of science and technology. Here, so to speak, the "fundamentalization" of a special education is observed. Moreover, the training of specialists in the newest branches of science and technology should be carried out only at institutes equipped with modern equipment and having efficient professors and instructors as well as scientists actively engaged in a particular field.

University or comprehensive special education (including a scientific degree for the working specialists) has become a quite serious problem. Apart from the programmes aimed at granting a scientific degree, shortened programmes on "interests", i.e. separate branches of knowledge, have become widespread (as an example, we can refer to the courses of public lectures delivered for engineers by the teachers of higher schools, in particular, on mathematics, physics, and electronics). This kind of instruction requires, first of all, improvement in the methods of representation, and is obviously primarily intended for acquiring knowledge in the field of fundamental science by working specialists.

In the USSR, it is believed that short-term, 2-3-week advanced courses for the managing staff of enterprises (directors, chief engineers, etc.) and 1-month special schools can be used for increasing qualifications. It is necessary to maintain the 1-year (or 2-year) courses for retraining specialists with a higher education; the instruction of these specialists would run parallel to their independent research work. This kind of instruction is completed by defending diploma work.

Local television should be more actively used for the purpose of retraining. For three or four terms, it is useful to organize the delivery and broadcast of lectures from institutes to audiences located within 80-100 km of their doors. Of top priority for the ideological outlook of the students is the study of social sciences.[3] One of the ways of *increasing the effectiveness of instruction in social sciences* is, in our opinion, the linking of the lectures with the orientation of specialization of the institute. Of course, this is important for all social sciences departments but it is especially so for that of philosophy. It is quite clear that the methodological function of philosophy takes account of the peculiarity of the institute. In lecture courses on dialectical and historical materialism, the teachers of the philosophy department at the MPhTI analyse, along with the general problems of philosophy, those philosophical problems of natural sciences; first of all, the branches in which the students are specializing. Not only does this contribute to a deeper study of natural sciences, but it is also a major factor for the formation of the students' scientific dialectico-materialistic ideology.

A link with special subjects in the instruction of social sciences is only one side of the process for improving the students' ideological outlook. The other side may be called *humanitarization* in the instruction of special, natural-scientific and technical subjects. Humanitarization has deep roots in integration processes occurring in modern science and in the processes of the blending of social, technical and natural sciences. Without attempting to elucidate all aspects of the questions, I would like to draw attention to three lines of humanitarization in the instruction of natural and historical subjects.

Firstly, the aspects associated with the development of scientific knowledge, especially with the establishment of scientific theories, must be reflected in the courses of lectures. Sometimes this viewpoint raises objections on the grounds that the students are in need of up-to-date knowledge and must have experience for its application in the solution of problems in a given branch of science. This opposition of historical and theoretico-applied aspects is hardly justified. The history of the construction of scientific theories permits us to look into the creative laboratory of scientists and to understand more fully the essence of their scientific results. Besides, such an approach increases the educational merits of the lectures since the students learn about advances in science in the socialist society and about the contribution made by its outstanding representatives.

Secondly, in lectures on general technical and applied branches of science it is sensible, along with special problems, also to discuss the social aspects of technical progress. Such questions as "man and technique", and "society and nature", must be reflected in the courses not only of social sciences but also special subjects.

Thirdly, fundamental sciences interact with both specific and general philosophical problems. This is especially typical of periods of scientific revolution when a conceptual reconstruction of science and revision of its numerous laws and principles take place, and which makes scientists discuss the most fundamental problems of science, including philosophical ones. This is the case in science itself and must be reflected in the instruction process.

Ideology is the most basic element of knowledge, that from which convictions are derived, and plays a significant role in influencing man's behaviour. The fundamentalization of education is fully reflected in an ideological background. The quality of the training of future specialists largely depends on the results obtained in this domain.

One of the most important questions for modern higher schools is the organization and development of research work. The principal task of science at the higher school level is the more productive utilization of the great scientific potential of higher schools and the involvement of the institutes with highly qualified specialists in performing major complex scientific investigations, applied developments, and so forth.

In the development of science at higher schools, a gradual transition from extensive forms of scientific activity to intensive ones will occur, i.e. further growth will be attained by increasing the effectiveness of scientific developments, rapid utilization of their results in industry, improving the system of management, increasing the level of the training of engineers and scientists, and so on. While the growth heretofore of the number of research workers and engineers has been proportional to the development of science and technology, in the future it will be difficult to maintain this rate of increase and the professional flexibility of researchers, i.e. their ability to rapidly master new directions of investigation, will play a more significant role. Higher education must provide for the development of top-priority fundamental investigations which nowadays are the basis for improving both the training of specialists and the theoretical level of applied developments. Of particular significance here is a close liaison between science and education, the training of specialists and scientific activity.

NOTES

1. National economy of the USSR in 1980. Stat. ezhegodnic. TsSU SSSR, M., Finance and Statistics, 1981.

2. O.M. Belotserkovskii, "Phystech looks into the future". *Zh. Nauka i Zhisn*, N2 (1974), 42-47. O.M. Belotserkoskii. "Technical universities". "Future of science", izd. *Znanie*, M., 7 (1974), 331-343.

3. O.M. Belotserkovskii and E.M. Chudinov. "Scientific-technological revolution and the training of specialists of a new type". *Zh. Politich. samoobrazovanie*, 6 (1980), 21-29.

Science as human need and the role of the mass media

S. Kapitza

In our modern age new scientific facts, discoveries in different fields of science, and inventions in technology are spread quickly and widely. Science not only serves as a satisfaction of human needs, but it is the primary human need in itself: the need of knowledge of the surrounding world. As the other human needs are satisfied, so the role of this aspiration towards scientific knowledge increases.

The popularizing of science and the dissemination of scientific knowledge has a long-standing and deep-rooted tradition within the cultural heritage of the USSR.

It is proper to consider the broader issues facing scientists, communicators of science, and the media. Perhaps from this high vantage point we can look around and see what we can aim for in our rapidly evolving world—a world in which people have attached so much hope to science. I shall mention only a few points that may be interesting to those present: personal impressions gained after some years' attempting to present science as an integral part of culture on a nationwide television network spanning ten time zones of the globe.

Today the results of science—new discoveries and inventions in technology—are all duly reported. The discovery of a new particle or a planet's satellite may even reach the scientific community through the mass media before the professional journals have had the opportunity of reporting the latest breakthrough. This modern deluge of information, although flowing at various levels, floods the layman and perplexes the scientist. To cope with this, we have made a concerted effort to popularize science, usually aiming at a particular audience, and relying on expert communicators, scientific journalists and scientists themselves. This is an important and ever-growing enterprise but, on the other hand, when you have an opportunity of capturing (or losing!) perhaps the largest television audience in the world, what should your approach be?

Our main role is to help people develop their *attitude* towards matters scientific. For this, we have to inform them of the facts of science. We have to popularize its concepts, but we should also excite their imagination, even entertain our audience, so as to make them think.

On every level what people really want is to understand what is bad or good, what is useful and what is necessary: what it all really means for them. These are universal questions that everyone has to face, and issues that everybody has to resolve for himself.

In a certain sense, each person is, if not a philosopher, a universalist—for these problems encompass all knowledge. With the sheer volume of things known, the vast realm of modern science, it is impossible to reach a general understanding by summing up all particulars. When on television, one cannot go deeply into specific problems. One has to generalize to an extent well beyond anything permissible in teaching. It should be emphasized that popular science, however useful and stimulating it may seem, is no substitute for education. In helping the public at large to develop attitudes towards science, we are doing what to a great extent is missing in much of teaching, where the broader issues of an education are sometimes lost. Yet the rapid change in our understanding of the world often exceeds the limit of even the best education and the scope of knowledge of the most educated person. Thus it is not the level of understanding that matters, but the general interest in things around us. It is to this interest we have to address ourselves.

What follows from this concept of developing attitudes among the public in general is that we have to engage the very best scientists to do so. Not to satisfy the vanity of the highbrow, but to instil trust in the message communicated. The most fascinating experience a television spokesman (who does not have a direct contact with his audience) can have is a chance encounter with his viewers. In the beginning, I was asked "Why invite a Nobel prize winner to discuss elementary issues, known to any professional?" It is not the special knowledge that matters; that is not so important. What we are after is the attitudes of these top people. After an initial period of some frustration, we have involved in the programme the finest brains in Soviet science, and have introduced to the public many first-rate scientists from all over the world. These personalities are best equipped to speak their minds. They have the scientific and moral authority to broadcast their attitudes, and this is what the public expects: the human message the public has a right to receive.

However important may seem the need to be, as they say, a good communicator—of mastering the tools of the media, the tricks of the trade—overriding everything is the sense of responsibility that we as scientists, as popularizers and expositors, have to face. The media, television most of all, have an uncanny, truly magic persuasive power. In whom, on matters of science, should this power be vested? Where resides the ultimate authority, or does it exist only in the depth of our personal consciousness? These are not easy or merely abstract questions. Take one example: for whom does an eminent scientist speak—for himself, for the scientific establishment, for world science? There is also the question of the social responsibilities of the media. The way, for example, nuclear energy and genetic engineering have been treated in a number of instances shows that the media are not beyond reproach. And even greater complications arise when socio-scientific issues are presented; for instance, the development of new pharmaceutical chemicals. Perhaps the most complex problems raised by science are the moral issues facing mankind: is there to be a new system of values in dealing with human problems and ethics that are to guide us? Here we again encounter profound and difficult questions that in every generation are facing the scientist, the philosopher and the layman.

Of course these are problems common to many societies. In engaging scientists from different countries with different intellectual traditions, we can and do contribute to an understanding between peoples.

Thus it is recognized today that science and scientists are essential links to establishing mutual understanding between nations.

Of all the world community's problems, I would like to point to the global problems. I am thinking of food and energy, the environment, resources and population, and last but not least the menace of arms. By their very nature, global problems are not only interdisciplinary but also international in scope. It has been said that in considering these problems we are paving the road towards a new way of thinking, a new mentality based on cooperation and parity, rather than sovereignty and superiority, as well as reinforcing policies of détente and coexistence.

In presenting global problems the consensus of opinion already formed by a number of international institutions (e.g. SIPRI, the International Institute for Applied Systems Analysis, the International Atomic Energy Agency, the Club of Rome, and the International Council of Scientific Unions) could help in expressing the point of view of science. It is essential to convey this information to the public. We cannot, moreover should not, expect this to be the final message. Guided by historical optimism and political realism on these imminent problems, we should strive to introduce scientific thinking, whose method, as every scientist knows, is often more important than the result itself. In these cases, we are also establishing a link between social and natural sciences that have long been separated by the dichotemy of the "two cultures".

The traffic of understanding carried by science is important, not only between the East and West where it helps in establishing mutual trust and a sense of security, but also in the North-South relationship. Here much more has to be done. We have to set up the necessary means for transferring and adapting all scientific matters matching this process with national educational levels in varying cultural traditions. We must not only bear in mind the immediate usefulness of science, such as applications in medicine or technology that are of paramount importance; we must also see that this pragmatic view does not overshadow the deeper human message in science—to understand things, not only to do.

The connection between science and culture, science and society, the necessity for a modern synthesis encompassing not only natural sciences but also those that are so aptly called in this part of the world, *la science morale*, is a pressing issue of our age. For many people, these high ideals of scientific humanism are associated with the noble mission of Unesco.

Microelectronics and society: report to the Club of Rome

A. King

This paper mainly refers to the recent Report to the Club of Rome, *For Better or for Worse*, devoted to the development of microelectronics and to its social consequences. An account of its development, together with a brief summary of the contents, may give an impression of at least one of the methods of operation of this international non-political group of individuals who seek to influence policy by encouraging debate on some of the major problems of our time.

The Club of Rome is a grouping of 100 individuals from some 40 countries, representing many disciplines and many shades of political opinion, but united in their concern for the future welfare of mankind. One of their main conceptual approaches relates to what they term "the world problematique", namely the tangled web of interacting problems which, in both policy and action terms, are generally considered separately and independently. Many of the contemporary problem areas, such as energy or increasing population, cannot be discussed in the vertical, sectoral terms of existing governmental structures, but sprawl untidily across nearly all the departments of a national administration. The Club of Rome endeavours to secure identification and understanding of the cross-linkages between problems and policies and to focus attention on aspects of the world problematique of growing importance which it feels are insufficiently appreciated or discussed. One of its methods, therefore, is to commission reports which, it is hoped, will be widely discussed and debated. It should be realized that these reports are *to* the Club and not *of* the Club; there is no attempt to reach a consensus between the members, but merely to draw the attention of policy-makers and the public to the impending importance of certain subjects.

Since this book is concerned essentially with methodological approaches to world problems, it may be useful to describe how this report on the consequences of microelectronic developments was organized. The subject was discussed at a plenary meeting of the Club, held in Berlin in 1979, where a panel of members and experts discussed the possible impact of this new technology on society, for a day. As a result, a study group was formed for discussion in depth and to undertake the preparation of a report, should this seem desirable. The convenors of this study group, who later became the editors of the report, were Professor Adam Schaff, the Polish sociologist, and Dr Gunter Friederichs of the Metalworkers Union of the Federal

Republic of Germany. The work of the study group was made possible by a grant from the Volkswagen Foundation of Germany.

Friederichs and Schaff began by constituting a balanced team of potential writers of the report, concerned with different facets, such as the nature of the technology as such, its range of applications, its economic impact, possible influences on the industrial enterprise and on the worker and his conditions of work, social consequences, microelectronics and war, effect on Third World development and general political considerations. Each of these eventually formed a chapter of the book. Two additional topics, microelectronics and energy and artificial intelligence, formed the basis of additional drafts, but were not, in the end, included. I was responsible for a general introductory chapter and for the final chapter of political considerations. Adam Schaff added a postscript of a philosophical nature on the concepts of work and occupation.

The group met on four occasions—for sessions of 3 or 4 days. The first was devoted to establishing the outline of the report and distributing the tasks. The authors then went home and wrote a preliminary version of their chapters, which were then subject to mutual scrutiny and criticism. Second drafts were then prepared and, at the third meeting of the group, were presented to a meeting in Hamburg of 28 experts, half of whom were from developing countries. This gave rise to a lively debate, during which many constructive criticisms and useful additional points were made, which were then taken into account by the authors during the days which followed the wider meeting. The authors then prepared their third drafts, which were edited by Friederichs and Schaff, with some assistance from myself as I had had the opportunity of discussing the topic widely in Europe, North America and Japan.

The title of the book, *For Better or for Worse*, indicated the uncertainties which surround the topic on microelectronic development and, especially its social consequences. In the past, despite the initial fears of the Luddites and Machine Wreckers at the beginning of the Industrial Revolution, technological development has always given rise in the end to more rather than less employment. Industrial technology has greatly lowered prices and increased wages so that the workers gradually became consumers and helped to create the tremendously increased markets for the products of industry. Many people therefore argue that the development of microelectronics will likewise create new industries, new markets and new jobs. Others are convinced that the microelectronics situation is utterly different from that of other technological developments; that it will lead gradually to an extensive degree of automation in industry as well as in the service sectors, involving greatly decreased demand for human labour. In this sense it will generate a new industrial, or more generally a new information revolution of which the industrial component will be one of many. A report of the National Academy of Sciences of Washington,[1] which takes this view, states: "the modern era of electronics has ushered in a second industrial revolution . . . its impact on society could be even greater than that of the original industrial revolution".

After lengthy debate the Club of Rome Group decided that the latter was the more probable of the two alternatives: that no development since the improvement and application of the steam engine by James Watt had such wide possibilities of impact across the whole spectrum of human activities, and that microelectronics was likely to precipitate fundamental changes in the structure of industry, the nature of work and eventually on the essentials of society. The debate continues and it is extremely important that the issues be discussed widely and on sound information. This is the justification and the importance of *For Better or for Worse*.

It may be useful here to summarise some of the salient features of microelectronic development. As recently as 30 years ago electronic devices such as the early computers worked by means of vacuum tubes or valves which were relatively bulky and energy-consuming. The first major innovation leading to *micro*electronics was the discovery of the transistor, a product of solid-state physics which used semi-conductor materials. The most important of these today are based on the non-metallic element silicon, to which traces of impurities such as boron or phosphorus are added, thus altering its electrical properties. Such semi-conductors function as miniature electronic switches, barring the flow of electrons or allowing them to pass. Being much smaller than the vacuum tube, the transistor quickly replaced the latter in all electronic equipment. However, transistors and other components had still to be wired together and, as a single piece of equipment might have thousands of such components, wiring was a delicate and costly process. Further miniaturization resulted in 1959 from the concept of the integrated circuit. Until then transistors were produced in the form of batches of thin wafers of silicon semi-conductor material, later to be wired together with other components. In the integrated circuit the connections were made directly on the silicon sliver, at first using very thin wire, but soon advancing to a process in which this was replaced by minute aluminium conductors deposited *in situ* on the silicon. The first integrated circuits were relatively simple, but as the techniques developed they rapidly became smaller and more complex. Today, closely packed integrated circuits may contain around 100,000 components on a chip measuring 5 mm across, and with the aluminium connections more than thirty times thinner than a human hair. As the Worldwatch paper on Microelectronics puts it: "in three decades, a whole roomful of vacuum tubes and other components has been reduced to the size of a cornflake".[2] But the process of refinement and miniaturization is by no means completed. It is confidently estimated that by the end of the present decade, chips containing at least a million elements will be available.

A further aspect of this remarkable development has been the sharp reduction in cost as a result of mass-production methods. In 1960 a single transistor cost about $10; today a transistor embedded in an integrated circuit costs a fraction of 1 cent.

The central feature of microelectronics development is the microprocessor, a special form of integrated circuit with functions of arithmetic logic and of control, similar to those of the central processing unit of a computer and contained on a single silicon chip. In addition, the microprocessor includes units to interpret instructions from the stored programme, to supply the controlled memory with the information necessary to retrieve instructions and send out data as required. The micro-computer, in contrast, is a complete system with memory, input-output, control and other chips in addition to the central microprocessor. Integrated circuitry makes it possible, in effect, to provide a brain and a memory of very small dimensions to any piece of equipment devised by man—and at a very moderate cost.

The extent of miniaturization is vividly illustrated by computer development. The first electronic digital computers, introduced at the end of the Second World War, were bulky installations containing as many as 7500 relays and switches, 18,000 vacuum tubes and 70,000 resistors. The next generation, that of the transistor, was considerably reduced in bulk and today's silicon chip generation is 300,000 times smaller, 10,000 times faster, much more efficient in its use of power and, at the same time, much more reliable. This trend is likely to continue further until the home computer is within the means of almost everyone.

While the rise of the microelectronics industry as such is important and fast-growing, especially significant at this time of economic recession, the use of its products and potentialities in other industries, and in the service sector, is still more significant. Its applications appear almost unlimited, with great possibilities for productivity increase and the elimination of dirty, dangerous and dull human work. For the first time, full or nearly full automation—not only of industrial processes, but of full production systems—becomes possible, technically and economically. Such automation would be difficult, if not impossible, to achieve by mechanical or macroelectronic means. It is certainly important for progressive countries to build up capacity for the manufacture of chips and microprocessors, not just for the economic benefits from such activities, but, more importantly as a means of gaining detailed and intimate knowledge of the techniques and of understanding better the potentialities for application. International diffusion of technology is sufficiently fast to enable all countries at a reasonably high level of industrialization to use and even to pioneer applications of the microprocessor and hence, in view of the wide range and competitive significance of these applications, it is essential for all countries wishing to avoid technological obsolescence to pursue these developments actively.

OPPORTUNITIES AND RISKS

The social and other consequences of the new information technology, including those connected with employment, will depend greatly on the speed of innovation. For example, very rapid automation of a particular industry could be initially disruptive, with considerable redundancy and insufficient time for retraining people for new tasks either within the enterprise or for alternative employment elsewhere. This could also mean that insufficient time is available for the evolution of new management techniques so as to permit efficient operation in radically changed circumstances. However, in the main, this is not likely to happen for many reasons, including those of capital scarcity and social resistance to change, though it could give rise to difficulties in particular sectors or plants.

We started from the hypothesis that the benefits which flow from the new technologies will appear to be so great that few of the industrialized countries will attempt to avoid or slow down the applications, particularly as they observe the advances in technology and in international trade which the pacemakers, Japan and the United States, will have made. If this is accepted, it will be necessary to explore in depth the probable nature and extent of the benefits, and to ascertain more clearly the probable social and cultural consequences on the basis of alternative scenarios and assuming different rates of introduction of the technologies. This will necessitate attempts to foresee consequences beyond the end of the present decade. If this is not done, and developments are planned merely on the basis of short-term rewards and immediate interests, it is inevitable that governments will attempt to absorb social and other consequences by marginal adjustment of existing policies and procedures. The result could only mean an increase in the rigidities and distortions as the century runs out, and a failure to shape the social transformations which the optimum use of the new information technologies will suggest. The fundamental question is, therefore, whether governments, with the support of informed public opinion, will be capable of using the new technologies deliberately and consciously to shape a better society, or merely and passively attempt to adjust to their consequences *post facto* as a matter of expediency.

The positive aspects of microelectronics development for humanity are enormous. The promise is that, through the ubiquitous application of the new techniques, work and living will be transformed to the benefit of the individual and of society. In particular, the microprocessor, in making possible a high degree of automation in industry and the services, should make it possible to provide the resources required by a country, including those for health education, food, welfare and defence, to provide a reasonable material standard of living and of security for everyone, with only a fraction of the physical work currently expended by humans and without depleting and degrading the environment. All this assumes, of course, that equitable means of distributing the wealth will be instituted.

Initially this may be seen in terms of the elimination of dirty, boring, repetitive and dangerous jobs, shorter hours and working lives, and hence more leisure. Later it could open the way to a society in which the individual would have the time, the resources and the opportunities for the achievement of personal fulfilment through the cultivation of his or her personal interests: artistic, scientific, craft, educational, sporting or otherwise.

It is impossible to foresee with any accuracy, at this early stage in the evolution of the information society, what will be the consequences for the individual and society as a whole. A number of trends are, however, discernible. The new technologies will, for instance tend to increase the interdependence and mutual knowledge of individuals and nations, through the immediate and comprehensive availability of information. They will also make for even greater complexity of institutions and societies, already difficult enough to manage. Industry will be considerably different from that of today, producing lighter and more compact products. The manufacturing centres will be cleaner than the traditional ones, with less need to cluster in large urban agglomerations. The nature and distribution of skills will be very different from those of today, and there will undoubtedly be heavy demands on the educational and training systems.

Since one feature of the new information and communication techniques is their capacity to annihilate distance, they may pave the way towards a fuller participation of the individual in public or corporate affairs and to all kinds of decentralization, which is a desideratum voiced in many societies today. This could well lead to the emergence of new systems of governance with more levels of decision-making than today, and of a dispersion of power. On the other hand, it has to be admitted that advanced electronic methods could bring to the unscrupulous or power-hungry leader the means to consolidate and wield absolute centralized power, through control of the activities and eventually perhaps even the thoughts of the population. Present concern for the privacy of data in a world where personal details are recorded and stored in data banks, whose computers talk ceaselessly to one another, is an initial and healthy recognition of this hopefully more distant danger.

A further preoccupation relates to the increasing vulnerability and fragility of societies ever more dependent on technological systems. Even at present levels of technology, the smooth functioning of cities and states depends upon technical devices which are exceedingly sensitive to disruption, either as a consequence of occasional malfunction or to sabotage and terrorism. This has already been dramatically shown with regard to nuclear incident, but even prolonged power cuts in cities such as New York are potentially dangerous. As human activity becomes more deeply computerized, such dangers could reach a higher order of magnitude. Power stations, nuclear reactors, oil refineries, communication networks and data banks all have their sensitive nerve centres, the destruction of which could disrupt the societies they serve. Such considerations should provide further and compelling arguments in favour of decentralization, and indeed the need to adopt policies of resilience rather than those which provide a rigid stability.

More subtle and difficult to discern, however, are the probable consequences of microelectronics on individuals living in the information society. In a culture dominated by sophisticated technology there is always a tendency towards a dichotomy between those who understand its nature and mechanisms and those who merely push buttons. Granted, it is not necessary to understand electronic theory to profit from television, but when the microprocessor has spawned to the extent that most of the equipment and artifacts of life take the form of little black boxes, and the level of sophistication of those who invent and design them and who devise the software has soared beyond the comprehension of the many, we could well be faced with a "two-cultures" situation, with a sharp distinction between the few who know and the many who do not. The emergence of a priesthood of technologists and technocrats is certainly undesirable, but by no means unavoidable. Its prevention lies in foreseeing trends of development, social and cultural as well as technological, in the level of general understanding and basically in the educational system.

More insidious still is perhaps the threat of isolation and eventually alienation that a microelectronic culture could generate. We are already familiar with the hypnotic attraction of the television screen; how it can chain people to their chairs by the hour, persuading children to enter the imagination of others rather than exercising their own. The cashless, jobless information society of the future could easily compound this effect. The concentration of many information inlets within a single room around the television screen and the family computer, providing information, education and entertainment options in great variety, facilities for household shopping and banking, electronic mail, and many other services, and coupled with much employment which will not necessitate movement from the home to the factory or office, could gradually immobilize the family and isolate it from rich human contact. This could lead to a creeping alienation of the individual; not the active counter-cultural alienation of today, but a passive and insidious alienation with leakage of human dignity and self-esteem. Of course this need not happen, but its possibility leads us, once more, to stress the need for prospective understanding of the social consequences of the information breakthrough, so that it may be used for the improvement of the human condition and not for its deterioration.

Space does not permit a discussion of many of the other concepts of *For Better or for Worse*. It contains an important but grim chapter on microelectronics and war, while the final section is devoted to a discussion of possible political consequences, mainly through changes in the pattern of international trade. It is suggested that, despite the present technological lead of the United States, Japan may well become the world pacemaker, because of her vigour in application of the techniques throughout industry and the unique qualities of this culture through the consensus approach and company paternalism to minimize resistance to change and absorb its consequences. Much consideration is also given to the probable impact of the microelectronics revolution on the various types of country within the Third World. It concludes that in the short term this new innovation will tend to weaken still further the economic position of such countries, especially those which have built up important markets in textiles, garment-making and the assembly of electronic equipment; the former are extremely labour-intensive and their advantage of cheap labour could be seriously eroded by the conversion of such traditional industries to highly automated science-based manufacturing in the advanced industrialized countries, while the assembly of electronic products is already a focus of automation.

The main message of the book is that the potential advantages of the micro-electronics revolution are so great that no country is likely to reject them. However, the social, cultural and other consequences will be radical, with many difficulties and dangers, but also with unprecedented promise for the transformation of society to the benefit of all mankind. Thus there is a quite major challenge to governments, industrial and social leaders and scientists to envisage the long-term promise of this new wave of technological revolution, to analyse its nature, potential and consequences in depth, and to exploit the new situation for the well-being of the peoples of all countries. To allow such development to evolve, without deliberate long-term planning, through motivations of immediate and narrowly national policies of expediency, would be to miss historic opportunities and to aggravate the difficulties.

NOTES

1. *Microstructures: Science, Engineering and Technology*, National Academy of Sciences, Washington, DC, 1979.
2. "Microelectronics at work; productivity and jobs in the World Economy", Worldwatch Paper 39, Washington, DC, October 1980.

The Global 2000 Report: a study in the scientific forecasting of human needs

P. M. O'Brien

INTRODUCTION

There are few, if any, other studies that I know of today that are as relevant for our consideration of "Scientific Forecasting and Human Needs" than the recently released *Global 2000 Report*. The Report provides an unusually comprehensive assessment of present and future human needs; it deals not only with conventional measures of human needs—such as food, energy, and mineral needs—but also with less conventional measures of environmental needs. The Report is also a study of scientific forecasting—its strengths, its weaknesses, the state of the art, and its application in the study of human needs.

With these two notions of the relevance of the *Global 2000 Report* in mind, I will review the Report's original mandate, summarizing its major conclusions regarding human needs, and outlining its implications for scientific forecasting.

REPORT MANDATE

On 23 May 1977, in his Environmental Message to the Congress, President Carter directed the Council on Environmental Quality and the Department of State to work together with other appropriate agencies of the US government to develop a study of probable changes in the world's population, natural resources, and environment through the end of the century. The study was to serve as the basis for longer-term government policy planning and evaluation. While originally commissioned as a 1-year study, the *Global 2000 Report*—as the study prepared by the Presidential Commission cam to be known—actually took almost 3 years to complete; and in addition to assessing probable population, resource, and environmental changes to the year 2000, it also assessed the US government's capacity to analyse and forecast change in these critical variables over time.

The study team ultimately included a small core staff which, in turn, drew on the Departments of Agriculture, Commerce, Energy, Interior, and State; the Central Intelligence Agency; the Executive Office of the President; the

Environmental Protection Agency; the National Aeronautics and Space Administration; and the National Science Foundation. The projection chapters of the Report were prepared by analysts from these federal agencies, which input and review from the core team and specialists from a number of other public and private organizations. Included among these organizations were the World Bank, the International Wheat and Maize Improvement Center, Brandeis University, Stanford University, the University of California at Davis, Case Western Reserve University, the Population Reference Bureau, the Rachel Carson Trust, and several dozen other organizations. The task of integrating the very diverse technical chapters that result fell to the core team. The study was released in the summer of 1979 by Study Director Gerald Barney.

Included in the Report were analyses of the current situation and prospects in the population, macroeconomic, climate, agricultural, fisheries, forestry, water, energy, fuel minerals, non-fuel minerals, and the environmental areas. In addition to these "substantive assessments", the Report contained a detailed documentation and evaluation of the models—however formal or informal they might be—underlying the study's projections and analyses. The result was a three-volume study—literally a short first volume of conclusions entitled *Entering the Twenty-First Century*, and two longer volumes on the sector-specific findings of the study teams and model documentation.

REPORT RESULTS: MEETING HUMAN NEEDS

The specific findings of the *Global 2000 Report* regarding human needs appear at first to vary quite widely, depending on the sector analyzed. The detail included in the study on topics as diverse as the future of tropical forests, the importance of dietary shifts towards livestock products, and changing fossil fuel technologies, are designed not for the general reader but for the technician. But underlying virtually all of the study's sectoral detail is a common finding that the world is likely, even assuming that technological progress continues to 2000 at the unprecedented rate experienced since 1950, to enjoy only a modest increase in human welfare at best. This relates, in the most simplistic terms, to the conflict between strong—frequently exponential—growth in demand for natural resources, including the environment and the finite, often unrenewable supply of these resources. The summary volume highlights what the chapter authors see in their individual sectoral assessments of this changing resource balance as the very real risk that population growth and environmental degradation may lead to actual decreases in human welfare in many parts of the world. The study teams also concluded that the pressure that meeting rising expectations will bring to bear on the earth's non-renewable and over-extended renewable resources between now and 2000 could undermine the world's biological capacity to sustain—let alone continue any improvement in—general welfare levels beyond 2000.

Adding to the gravity of the problem the Report sees as the world approaches 2000 is what the study core team identified was the possibility of an optimistic bias in the individual projection chapters due to the lack of cross-sectoral resource linkages. The Report's conclusions were generally built up from discrete, often unlinked models designed for use by sector-specific decision-makers. The models all tended to treat cross-sectoral resource links as exogeneous; this tendency allowed the energy, food, minerals, etc. study teams to "assume" that the land, labour, and capital needed to fuel their sectors were readily available from outside their sectors.

In short, for example, there was no investment capital model to rationalize and—where appropriate—constrain the different sector models' demand for capital. The few highly aggregate models developed to analyse cross-sectoral resource damand—albeit without the detail necessary to make sector-specific management decisions—suggests the world's resource-carrying capacity may be more limited, and the outlook for welfare improvements less optimistic, than the Report's simple projection add-up suggests.

This basic finding depends in part on the study's concentration on a no-policy-change scenario. In fact, changes in government policies are identified throughout the Report as the means of balancing the increasing demand for basic resources—both natural and man-made—associated with meeting human needs and the finite—in some cases even eroding—supply of these resources. The study also cites technology as a means of redressing the resource imbalance—particularly technologies geared to increasing sustainable resource productivity levels. The Report sees increasing dependence on record-breaking advances in fossil fuel technologies that increase resource productivity in the short term, but at the cost of sustainable resource productivity in the long run, as counterproductive. At best, dependence on inappropriate technologies can increase the world's supplies of food, clothing, and shelter significantly in the short term of one or two decades; at worst, however, it can involve transforming many "renewable" activities into non-renewable activities that ultimately reduce the world's carrying capacity. The classical example here is the gradual transformation of a renewable agriculture dependent on the management of land, water, and labour into an increasingly non-renewable activity often dependent on mining land and water resources and using fossil fuels.

These conclusions are disturbing enough for countries such as the United States and the Soviet Union where basic human needs are met and the welfare gains looked for over the next two decades can well be characterized as luxuries. They are far more disconcerting for the developing countries, however; where they translate into increased difficulty sustaining large numbers of people on the edge of debilitating malnutrition, with minimal medical services, and with the most limited educational or employment opportunities.

REPORT RESULTS: THE IMPORTANCE OF SCIENTIFIC FORECASTING

Perhaps the most relevant finding of the *Global Food Report* for us here today, aside from its resource balance findings, is the Report's notion of the growing importance of scientific forecasting and its use in policy analysis. This finding follows directly from the Report's conclusion that improved resource development and management decisions by governments can: (1) improve the human welfare levels possible in a particular resource setting in the short term of decades; and (2) possibly even change the resource setting itself over the longer term.

As the individual study teams were quick to point out, scientific forecasting and policy analysis are not new to sector-specific resource decision-making. Virtually all of the study team personnel were directly involved in planning and evaluating resource development and allocation in their home agencies. The study teams were quick to realize, however, that improved scientific forecasting and policy analysis are becoming even more critical in planning resource interactions between sectors.

But the study also makes a convincing case that the world's scientific forecasting capacity and its ancillary policy analysis capability are currently not up to the job they face over the next 20-30 years. However, a number of developments over the last decade suggest the analytic underpinning needed to improve the world's resource management can be developed over the decade ahead.

On the positive side, the study notes that tremendous technical advances in computer science, mathematics, and modelling—i.e. the technologies underlying scientific forecasting—have been made over the last 20 years. Sector-specific modelling at the start of the 1980s—whether it be based on input-output analysis, econometrics, biometrics or a host of other variants—is virtually unrecognizable to even the most advanced analysts at work in the 1950s and early 1960s. Moreover, the information base supporting sector modelling has also improved significantly. Contrary to the information systems of the past, the information systems in place in the advanced countries can support increasingly sophisticated research to identify and quantify key technical and behavioural relationships. Perhaps the most striking advance is the modeller's ability to combine disciplines and, crossing sectoral lines, to do more comprehensive analysis. Hence, while the concerns President Carter voiced in 1977 about the longer-range resource balance did not differ significantly from the concerns voiced a half a dozen Presidents before him, our capacity to analyse that resources balance, and evaluate the policies affecting it, is significantly better.

On the negative side, while there is an increased realization in both government and private circles that resource problems are seldom sector- or discipline-specific, progress towards actually developing cross-sectoral, interdisciplinary analysis has been limited. The best scientific forecasting and policy analysis currently being done involves input from a range of disciplines. In agriculture, for example, this involves cooperation between plant scientists, soil and water scientists, and agricultural economists working to model the farm and food sectors. The best work currently being done in the water management area, for example, involves agriculturalists, city planners, and climatologists as well as hydrologists. But these interdisciplinary, cross-sector activities are still the exception rather than the rule.

This relates directly to the economics involved in developing and using the necessarily complex, somewhat academic, models involved. For example, funding for agricultural modelling in the United States comes largely from the Department of Agriculture and the land grant universities—both of which are traditionally far more interested in farm and food sector decision-making and impact analyses than cross-sectoral analyses. Moreover, even should their interests include cross-sectoral analyses, their input into decision-making outside agriculture is limited at best. Hence, until both the funding for development and a decision-maker clietele is found, the gains in resource management possible from interdisciplinary, cross-sectoral analyses will continue to be potential rather than actual.

Even more pronounced than this lag in interdisciplinary, cross-sectoral application has been that in applying scientific forecasting to cross-country problems. It is precisely this area of interdisciplinary, cross-sectoral, cross-country planning of resource development and use that the *Global 2000 Report* sees as critical. But the obstacles to such activity are considerable. For a start, the information basis available for modelling resource activity in many of the developing countries is virtually non-existent; forecasting technology in the most advanced developed countries has clearly leapt well ahead of data collection in the rest of the world.

What makes this information gap particularly troublesome is the Report's contention that the resource balance in 2000 and beyond could well be tightest in the developing countries with the most limited information bases. An equally serious problem is what many model developers—and even more so the interests funding model development—see as the limited payoff on global modelling. While the problems the *Global 2000 Report* foresees are truly global in perspective, resource policy decisions are now—and will quite likely continue beyond 2000 to be—national policy decisions at best.

The study team pointed these two problems out quite graphically, if somewhat inadvertently. Regarding the need for interdisciplinary and cross-sector analyses, critical aspects of the food and agriculture problem are treated using no less than seven different models—the population model, the food model, the GNP model, the water modelling, analysis of tropical forestry problems, the fisheries analysis, and the environmental impact analyses. The study teams were forced to link these seven pieces in a somewhat haphazard manner that made impact assessment or simulation cumbersome at best for use in even farm sector-specific resource decision-making. The results of the study's use of the few highly aggregated, interdisciplinary and cross-sectoral models to remedy this shortcoming point up one of the most serious dimensions of the resource problem—the competition between sectors for increasingly scarce "common" resources which, if not provided for, can seriously bias study results.

Regarding the need for cross-country analysis, the study is forced to note that, in even those few modelling activities that have international components, the models generally include a full-scale US model, possibly a reasonably complete Western European and Japanese component, but only the most limited quantitative notion of how the rest of the world works.

CONCLUSION

In conclusion, let me re-emphasize three points. The *Global 2000 Report* makes a rather convincing case that meeting basic human needs will be the most challenging problem we face ahead. Meeting this challenge will depend in a very real way on our use of scientific forecasting and policy analysis as the basis for better resource development and management. Lastly, considerably more work is needed on our part as forecasters to develop an interdisciplinary, cross-sectoral, cross-country perspective if we are to meet the challenge ahead.

Summary Discussion

O. Zakharova

In the discussion and interventions the following issues were touched upon:

(1) general methodological problems of scientific/technological forecasting and of the study of the dynamics of human needs;

(2) examination from this point of view of various projects and programmes, both national and international;

(3) examination of certain modern trends in scientific/technological development from the point of view of their impact on man and society.

The main discussions on the methodological problems followed the presentations made by Acad. J. Gvishiani (USSR), Acad. I. Malecki (Poland), Prof. D. Meadows (USA) and Prof. H. Bossel (Federal Republic of Germany). The following participants took part in the discussion: Prof. R. Ayres (USA), Acad. E. Andronicashvili (USSR), Acad. O. Bielotserkovski (USSR), Acad. S. Belyaev (USSR), Acad. E. Fiodorov (USSR), Acad. J. Gvishiani (USSR), Prof. K. Holling (IIASA), Dr I. de Hemptinne (Unesco), Acad. L. Kantorovich (USSR), Prof. S. Kapitza (USSR), as well as members of the Institute for Systems Studies: Dr V. Gelovani, Dr V. Klokov, Prof. N. Lapin, Prof. O. Laritchev, Prof. B. Milner, Prof. I. Novik, Prof. S. Shatalin and others. They stressed the necessity of close international cooperation in this important field, and of organizing international projects. They also emphasized the need for a multidisciplinary approach and interdisciplinary methods indispensable for a comprehensive vision of the complex objective. Representing one such method, the systems approach provides modern scientists with an effective tool which has been more and more widely used over recent years, not only in scientific/technological forecasting but also in the study of human and societal needs and their dynamics. The importance of further elaboration of systems methods and their application to the problems of scientific forecasting in relation to human needs was especially stressed by Acad. J. Gvishiani, Prof. K. Holling and other participants. Special emphasis was laid on the choice of adequate techniques, the correct combination of formal and non-formal, and quantitative and qualitative methods and a substantial, in-depth interpretation of the results and data obtained.

Referring to the experience gained by the developers of the Bariloche World

Model, Prof. C. Mallman from the Bariloche Foundation emphasized the importance of a human needs approach in scientific research and in scientific forecasting exercises. Being an essentially intellectual and rational enterprise, he pointed out, science cannot satisfy some of the more fundamental human needs, whereas science grounded on a human needs approach could make a substantially greater contribution towards the resolution of human and social problems than that which science makes today. At the same time, Prof. C. Mallman warned of an excessively pragmatic orientation of science. Human needs criteria as a policy guideline should be followed with respect for the scientific endeavour, and not used to destroy the scientist's freedom and creativity.

The participants of the symposium were unanimous in pointing out that the most urgent of collective human needs, equally important for all people all over the world, is the need for peace and prevention of wars. A lasting peace is not only a prerequisite for the survival of mankind and for the full realization of human potentials, but also a necessary precondition for the mobilization of the scientific potential of the planet for the solution of the most crucial social problems and the satisfaction of material and spiritual human needs. A militarist orientation of scientific development represents not only the meaningless misuse of human creativity, but also one of the main obstacles to the harmonization of the interrelationship between Science and Man. It was unanimously agreed that the struggle for peace had become the most important social responsibility of scientists in the contemporary world.

As was stressed by Acad. J. Gvishiani, human needs form an historical category and as such are exposed to changes both in content and form under various social, national and cultural conditions. Hence, one of the basic factors affecting human needs is the influence on their formation and the hierarchy of differing sociocultural factors and moral-ethical attitudes. Stressing in this connection the importance of the interdependence between human needs and changes in moral settings, A. King gave the example of the "consumer-society" attitude to nature which is widespread in some western countries. Dr G. Stetzenko noted that the structure of human needs strongly depends on concrete social conditions in a given society, and stressed the importance of creating a suitable social climate for the forming and satisfaction of material and spiritual human needs, which would favour the fullest realization of human potential.

Prof. M. Hogbe-Nlend, President of the African Mathematical Union, noting the African peoples' urgent need for development, both at the individual and collective levels, emphasized that this can no longer be considered simply in terms of economic growth, but must take into account the need for African cultural identity. Referring to the role of science in solving these problems, Prof. M. Hogbe-Nlend emphasized that there are two fundamental ways of putting science and technology to the service of development: firstly, to create, reassess, develop and use indigenous scientific and technological knowledge; and, secondly, to master non-African science and technology so as to assimilate them as opposed to being assimilated by them. Africa, he pointed out, must follow both routes simultaneously.

Referring to methodological problems, Prof. R. Ayres (USA) stressed the importance of "sharpening" the question of human needs, concentrating research efforts on the most crucial aspects of human needs dynamics, taking into account the fact that there are many different kinds of "needs", both short- and long-term. In fact, they represent, noted Prof. R. Ayres, a hierarchical system at the highest level of which there are needs connected with survival (food, shelter, reproduction), then wealth, freedom, justice, knowledge, human potential, etc. Even at the highest level, there are inconsistencies between short-term needs (food, shelter) and long-term (e.g. reproduction, survival of the race, etc.): for instance, population growth interferes with the ability to exist. Research on human needs must be concentrated on these inconsistencies and incompatibilities. He gave an example of the use of simplified models describing the interdependence between mankind, the animal world and the green biomass and demonstrating the need for operating and repairing mechanisms for maintaining a non-zero balance.

Taking into account methodological and ethical problems linked with human needs research, all those who took part in the discussion were unanimous in stressing that any attempt to intervene, influence and regulate these fragile mechanisms demands the utmost discretion and implies great responsibility. This responsibility was especially stressed by Dr Y. de Hemptinne, who warned of the possibly dangerous, and probably irreparable, consequences of careless interference with these delicate and spontaneous mechanisms.

S. Shatalin pointed out that in studying the processes of the forming and satisfaction of human needs, one has to take into account the full extent of their spontaneity and be clearly aware of the extent to which these processes can be controlled and managed.

Accordingly, research activities must focus on the methodological problems relating to the formation, dynamics and pattern of human needs. The extreme complexity and variability of the systems under study pose serious, and so far unresolved, methodological problems. A cautious and highly self-critical attitude to the obtained results must be accompanied by a sincere recognition of the fact that it is dangerous to study complex, real-life processes by simple or simplified methods. What is needed is a continuous correction and enrichment of the methods used, more extensive studies of a conceptual character, and the application of their results to practical forecasting.

Investigation of public opinion, noted Prof. S. Kapitza, can efficiently contribute to the better understanding of these processes and mechanisms. Special emphasis is laid in present-day public opinion on science and its benefits and dangers for man and society. Stressing the importance of concentrating research efforts on problems related to the satisfaction of the most urgent human needs, Prof. S. Kapitza at the same time warned against an excessively utilitarian, instrumental attitude to science and its role in society. Science is not only a very powerful means of solving crucial problems in our contemporary world, but also represents one of the most ancient human needs; the need to know and to understand. As the needs related to survival come to be more and more satisfied, the role of this spiritual need will increase, since it contributes considerably to the self-realization of the individual. Contrary to material needs, spiritual ones, such as art, culture, science, etc., are inherently non-saturable and their development can help weaken the dangerous tendency towards unlimited and non-satisfying consumerism.

As a drive towards unlimited consumption, consumerism poses very serious problems and requires reorientation and a revision of societal values.

The issue of the role of science as a human need is closely related to the balance between fundamental and applied science in science policy. The core of scientific and technological forecasting, with respect to human needs, is the predictability of the very development of science. A discussion on this issue centred around the presentation of Acad. I. Malecki (Poland), who touched upon the assessment of trends in the development of scientific knowledge and the possibility of the practical utilization of estimates. Emphasizing the significance of fundamental research as a potential scientific "bank" for humanity, and its own value, irrespective of the present-day practical application, the speakers stressed that a narrow, pragmatic approach to theoretical results cannot be justified. As was pointed out by Acad. Malecki, nowadays fundamental science directly contributes to the satisfaction of human needs but cannot be evaluated only on this basis. Touching upon the role of fundamental research in scientific and technological development, Acad. D. Belyajev (USSR) pointed out that, in forecasting scientific and technological progress and determining the balance between basic and applied research, it is necessary to account for contingency, unpredictability, and the impossibility of foreseeing not only the practical application of the research outcome but the results themselves. Unpredictability, he said, was and is the driving force of the advancement of science. A certain part of scientific effort and expenditure should always be devoted to scientific research not linked with any immediate practical purpose, noted Acad. Belayev. When dealing with certain spheres of fundamental research one cannot predict the possible future application of its results, although the effort to evaluate its possible social consequences has always been a major social responsibility among scientists.

The impact which science can make on the satisfaction of human needs and the attainment of societal goals depends considerably on the structure of training of scientific personnel and higher education, noted Acad. O. Bielotserkovski (USSR), stressing the importance of the humanization of education in the natural sciences. He noted that, due to the ever-growing role of science in the solution of social problems, it is increasingly important to forecast the demand for scientific personnel and their skills in line with the problems that science would face tomorrow. All this, he emphasized, poses serious problems to the higher education system, which must be oriented towards the future, since present-day students are the potential scientists and scholars of tomorrow. He placed special emphasis on promoting a wide interdisciplinary outlook among specialists, and a good command of systems techniques to ensure adequate research on these problems.

Discussing the problems of the training of scientists, Acad. J. Gvishiani, Prof. D. Meadows and others emphasized that scientists need to have a comprehensive, systems approach to today's complex problems. Systems methods should therefore become a part of university programmes.

Science brings very rapid and unexpected changes to human life, transforming both living conditions and the environment. The problem of adaptation to these changes was touched upon by Prof. K. Holling, Prof. A. King, Prof. D. Meadows, Acad. J. Gvishiani, and other participants. Acknowledging the shortcomings and imperfections of existing techniques in the face of complex current problems, Prof. Meadows, nevertheless, called for their wider application in order to improve the validity of decisions. As was stressed by Prof. Meadows, one of the most important social functions of scientific forecasting is the timely exposure of these future changes, thus giving people adequate warning of their possible consequences, both positive and negative. The ever-increasing tempo of changes causes feelings of fear and instability in man, added to uncertainty when facing the future and helplessness when faced with unexpected events threatening both himself and his relatives. Adequate information based on sound knowledge and the results of scientific forecasting is a major element in the building of mechanisms for the flexible adaptation of man to change.

The now late Acad. E.K. Fiodorov called for extensive activities on a global scale aimed at the prevention of anthropogenic impact on the environment. He pointed out that in favourable sociopolitical conditions the capacities of mankind for solution of pressing complex problems are limitless.

The significance of research on the problems of human adaptation to the rapid changes brought about by scientific and technological progress was emphasized by K. Holling, Director of IIASA. Research into the means of providing the necessary resilience and adaptability in social structures and institutions for these changes, he noted, must be a priority for systems analysts, along with the traditional problems such as improvement of forecasting and management.

The study of the problems of adaptation by man, social institutions and human society as a whole to the changes brought about by the development of science and its applications represents a very serious challenge to science and scientists, and especially to those dealing with interdisciplinary methods and systems approaches, pointed out Prof. K. Holling.

Referring to earlier interventions by Acad. E. Fiodorov and Prof. D. Meadows, Prof. Holling noted that, while presenting sharply contrasting forecasts of the future, each of them had developed a consistent chain of causative arguments, leading in one case to economic and material growth and in another to a limited material growth. Acad. E. Fiodorov emphasized man's demonstrable ingenuity in transforming material power from one use to another.
Prof. Meadows saw a fundamental limit in the carrying capacities of air, land and water to support development. Therein, noted Prof. Holling, lies the dilemma for forecasting models, which are hypotheses or clusters of interrelated hypotheses, but, unlike those studied within the framework of traditional science, they are not subject to disproof. The scale and magnitude of their prediction exceed any realistic capacity to invalidate the models, or to put them at risk. Hence the only limit to the growth of forecasting models of this type in the ingenuity of the human mind.

The tradition of systems analysis is to place great emphasis on finding better ways of predicting and regulating. But we now encounter inherently unforeseeable events covering a bewildering range, and threats to survival of a frightening magnitude. Why is there such a demand for such visions of hope or of doom? I believe, noted Prof. K. Holling, that it is because people have lost their sense of certitude—have lost their sense that, whatever unexpected events occur, they will have the ability to adapt. Hence, in addition to doing the best we can in our efforts to predict and regulate, equal effort should be directed towards finding ways of increasing the adaptive capability of individuals, institutions and societies (for example, in agriculture, resource development and energy design; and in technology and social enterprise there is need for not just efficient but also resilient design).

Prediction, regulation and adaptation are like three legs of a stool. All are needed for balance, noted Prof. K. Holling, emphasizing the need to explore different concepts and perceptions of surprise, crisis and change and to compare and contrast a range of case studies of surprise-ridden management and development. The goal is to identify the different categories of surprise generated by the intersection of the natural environment, technology and society and to define ingredients for adaptive behaviour and management. These studies would be an attempt by the scientific community to respond to the sense of urgency concerning the vulnerability of institutions, nations and mankind to unexpected events and surprises that emerge from the intersection of society, natural environment and technology. This atmosphere of crisis, noted Prof. Holling, generates extremes and hostility to the values of the mind found in science, scholarship and technology; and threats to the physical, intellectual and creative spirit of man. An in-depth examination of the above problems demands the cooperation of scientists experienced in the scientific, technological and societal aspects of change, and requires the integration of concepts, methods and people with comprehensive techniques and the systems approach.

The importance of improved forecasting techniques as applied to the problems of the adaptation of individuals, social institutions, and society at large to forthcoming changes in the social sphere was also stressed by Dr A. King, President of IFIAS, who called for extending international cooperation in this field. He noted that all the crises confronting contemporary societies, which bear witness to the instabilities inherent in their development and to the lack of adequate adaptation mechanisms, demonstrate clearly that science and technology alone cannot provide stable social progress. The rapid tempo of change renders impossible and futile all attempts to attain any measure of stability or durability. The goal to reach is not this lost stability but a dynamic equilibrium based on the flexibility of society and its capacity for continuous adaptation to constantly changing conditions. All this emphasizes the urgent need for scientific instruments, methods, techniques and approaches which could provide a basis for such adaptation to a rapidly changing world. The comprehensive study of these problems can only be accomplished by large multidisciplinary scientific teams, drawing deeply on the experience accumulated in various countries. Hence, close international cooperation in this field becomes an imperative of our time.

The growing role which science and technology play in modern societies increases the importance of planning in scientific development. Together with efforts aimed at restructuring scientific research according to these new challenges, special attention should be paid to promoting constructive interdisciplinary cooperation, especially among young scientists. The contemporary stage in the development of science, noted Dr King, can in many senses be considered as a period of transition, which imposes the necessity of new approaches and new solutions.

The importance of a correct methodological approach to the assessment of the effectiveness of technological progress, and to the choice of criteria for such an estimation, was stressed by Acad. L.V. Kantorovich (USSR) who pointed out the negative consequences of science and technology's orientation towards military purposes.

The most lively discussions centred around the methodological problems of forecasting, relating to the effectiveness and adequacy of the methods used.

Special emphasis was placed on modelling as the most developed and widely accepted technique for forecasting socioeconomic processes on the national, regional and global scale.

The construction of models reflecting the properties of the system under study and its goals (noted Acad. J. Gvishiani) is one of the traditional principles of the systems approach. Models reflecting a certain holistic view of the object and the relationships of its major subsystems and components, are always based on a certain concept reflecting the world outlook of the researchers. All models are implicitly normative and conditioned by the sociopsychological milieu. Explicit acknowledgment of this by those who develop and use models is essential if misunderstanding is to be avoided. So far, the models have mainly concentrated on economic indicators as an objective or yardstick. Very little attention has been given to humanitarian and social, as against economic, indicators. In the course of the Symposium's discussions it also appeared that some considerations of overriding importance had been left aside in models of the world's future developed over the last decade, such as political and social tensions or conflicts, the problems of disarmament and its influence on the solution of global problems, issues of global climatic variations, etc. Disarmament, noted participants, would open new opportunities due to the emergence of extra resources presently employed in the arms race and in maintaining huge armed forces. The above opportunities have to be studied both from the point of view of the need for resources and the possibility of their release. On the other hand, curbing of the arms race will obviously contribute to harnessing inflation and other problems.

At the same time, it was noted that, as the concept of global modelling is developing, more emphasis is being put on the data and scenarios representing core problems of model exercises. Already, there are problems for which the use of models is justified and effective, and can provide good results for decisions and the solution of practical questions. The experience gained by many countries—particularly the USSR and the USA, and by organizations such as the EEC, OECD, etc.—in the application of this technique to the study of a number of problems provides solid ground for treating these methods as rather promising for the prediction of a certain class of processes. In particular, they seem very effective for the study of global problems. As was stressed by Prof. D. Meadows, the problem for the time being lies not so much in further improving these methods, but rather in developing practical ways and channels for their wide implementation in the decision-making process. What is important here is to convince decision-makers of the usefulness of models and of their indispensability when dealing with complex contemporary problems, noted Prof. D. Meadows. Greater communication between methodologists, modellers and users, including social scientists, would be beneficial in this regard. The essential point in communication with the users is to explain very clearly, explicitly and without mystification the hypotheses and shortcomings of the models.

At the same time, it was pointed out, the effectiveness of this technique is largely dependent on whether, and to what extent, it is based on scientifically valid studies of a conceptual nature, supplemented by qualitative techniques. While employing this method one must always be careful to limit the simplification, formalization and quantification of the processes. One should seek to combine qualitative and quantitative interpretations with the systematization of qualitative characteristics.

When employing simulation techniques it is important to have a true systems approach to the object under study, i.e. consideration of the entire range of essential and relevant factors of its development with respect to the formulated objectives. Otherwise, where the original assumptions and data are one-sided, unsystematic or perhaps biased, the employment of mathematical models—however perfect the hardware is—will be more like the "window dressing" of scientific objectivity and, in fact, will result in rather trivial explanations and often meaningless recommendations.

The mathematical modelling of processes involved in socioeconomic systems poses a host of problems determined by the properties of the systems themselves. The most serious one arises due to the lack of detailed quantitative data concerning the laws of the development process. As a result, there can be no single model of socioeconomic development for the solution of a wide range of problems. It was noted that, when studying the processes of global development, it is desirable to build a special model for each specific purpose (as is done for the majority of systems problems) which results from interdisciplinary research in different areas of social, natural and engineering sciences.

Some of the participants, stressing the imperfections of modelling techniques used up to the present, were doubtful as to their wide implementation in decision-making, calling for more active utilization of expert assessments and respective methods. As was noted by Prof. O. Laritchev (USSR), these methods are shown to be efficient when dealing with problems concerning the elaboration of science policy and its priorities. Both in the course of discussions of methodological problems and of such specific international projects as Interfutures or FAST, the participants unanimously noted that large-scale models have no prospects. They spoke out for the application of the more focused and clearly oriented models developed for specific research purposes. However, the study of real-life socioeconomic processes with the help of a family or families of such models runs into the problem of their combination, integration and compatibility within the frameworks of confined research activities. In particular, as was noted by Dr P. O'Brien (USA), this obstacle could not be overcome by the authors of the American project Global 2000.

An important factor considerably affecting the quality of results, is the level of model disaggregation, i.e. the degree of detail in the process description. One would think that the more information used, the more correct the results. However, by increasing the number of parameters which describe the behaviour of a real-life system, and accordingly complicate the model, one risks crossing the so-called "interpretation threshold", i.e. losing the ability to adequately perceive and evaluate the results.

On the other hand, the utilization of highly aggregated models is fraught with serious errors of a different kind: subconscious or conscious "projection into the future" of those trends, structures, interrelations and ideas, needs and values that seem habitual in the present-day world. These "paradigms of the present", built into the highly aggregated models, inevitably distort the picture of the future, for the development, as it is known, is not confined to purely quantitative changes but implies deep rearrangement of the structure itself and the interrelations between individual socioeconomic processes; and it is these processes which play the determining role here.

As was mentioned in the course of the discussion, in order to be able to adequately reflect the trends of the socioeconomic system's development, a system of models must reflect every significant process, especially with reference to the future. Not all of them yield to strict formalization. For example, the tools of economic process simulation are developed sufficiently well, while the sociopolitical processes are formalized to a limited degree. Attempts to formalize and simulate processes of this kind in the long term sometimes lead to serious errors in conclusions and predictions. This is due to the fact that they are either not formalized at all, or do not yield at present, for some reason or other, to formalization.

During the discussion many criticisms were made concerning the models and techniques so far used in scientific/technological forecasting and studies in the field of human needs. It was pointed out that global modelling should develop on a multidisciplinary basis and incorporate new, relevant results in different fields and subjects. A worrying fact about global models is that due to factors such as their complexity, their multidisciplinary content, the lack of suitable documentation, etc., many people willing to follow these activities are led to base their opinions on inadequate and sometimes misleading information. In this respect those who took part in the discussions emphasized the need for special efforts to popularize modelling amongst the general public.

The solution of practical tasks through modelling requires reliable informational support and building of respective data banks. As was underlined by participants, it is the reliability of data on the process studies which, to a considerable degree, determines the possibility and validity of model aggregation. Besides, it is very important that the simulation techniques be supplemented with flexible scenarios providing opportunities for a substantial increase of the range of the problems under study.

The discussions revealed that at present there is some discrepancy in the present situation. On the one hand, the ever-growing dynamism of scientific and technological development and its social impact urges scientists to develop reliable forecasting tools effective in decision-making. On the other hand, the complexity of modern socioeconomic processes imposes—at least with the current level of knowledge—objective constraints on the practical application of forecasts. A scientist, fully realizing his social responsibility towards the results obtained, faces a complex dilemma: the practice urges him to produce scientifically validated forecasting techniques. At the same time, the very practical orientation of research, implying some additional responsibility, makes him deal more carefully and critically with the conclusions obtained by these methods. Evidently, one of the ways out is to seek more vigorously the services of modellers directly involved in making and implementing perspective decisions, for the very process of research in the field of scientific and technological forecasts, and formulation of objectives and interpretation of results. Such "bridges" between science and decision-makers would enrich both the science of forecasting and the decision-making practice. Covering and taking into account the entire variety of processes and factors, essential for the solution of a certain problem, this tool must at the same time be relatively simple and easy to use by non-specialists.

It is this perception of research objectives that was realized in a concrete project, "Modelling of global development", which has been carried out at the All-Union Research Institute for Systems Studies (VNIISI) since 1977. The leading researchers of the Institute, Dr V. Gelovani, Dr V. Klokov, and Dr N. Lapin, described their work on the project. The latter is aimed not at construction of a single global model, simulating the development of the world and somehow or other fixing the interdependence of its sub-systems, but at a *simulation system*. It involves a system of global development models, a data bank, an interpretation block and, finally, a dialogue system allowing interaction with the machine. This structure of the modelling system makes it possible to continually increase the latter's capacities by varying, depending on the nature of the task, the number of models in use, expanding the data bank, improving the scenarios. As was noted in the interventions relating to this, one of the advantages of the system is that it overcomes the common problems often arising from the interaction with computers, i.e. either a "machine fetishism", or the development of a "computer inferiority complex". The sophisticated interaction system can be easily operated by any expert having no command of the specialized machine language or the translation procedure. The design of this simulation system required a number of new and daring scientific and technological solutions. Some things have been successfully accomplished; others are on the way to implementation. There are, however, noted the developers, problems which currently remain unsolved. At the same time, a host of intricate problems can be successfully solved today on this modelling system, said Dr V. Gelovani.

Summing up the discussion on the methodological problems of forecasting and modelling, Acad. J. Gvishiani emphasized the special significance attached to the problem of the choice of adequate techniques. This, emphasized Acad. J. Gvishiani, is especially important in the context of the Symposium, since scientific forecasting as well as the forecasting of its impact on human needs—both material and spiritual—and on the means for their satisfaction, encounter complex methodological problems and do not yield to simple and simplified methods of research. Modelling techniques should be applied rather carefully. One should not overlook the problem of the adequacy of the applied modelling techniques. In general, when applying these methods, we should constantly remember that existing modelling techniques and the models themselves are not yet adequate for, or consistent with, the complexity and diversity of complex modern processes, and cannot reflect this complexity. This does not mean, of course, that we must completely discard these techniques, noted Acad. J. Gvishiani.

Hence, it is especially important to unite the efforts of scientists working in different countries and research institutions applying and formulating models of different types and for different applications.

The current stage of model-building, noted Acad. J. Gvishiani, is mostly distinguished by the empirical approach. We are unable, as yet, to combine the available empirical experience and the theoretical approaches. We must strive to bridge the gap in theoretical studies on the one hand, and empirical research on the other. Special emphasis here should be placed on informational support and data banks, since only they enable us to apply modelling techniques to the study of a definite class of processes, even where the processes yield in principle to such studies, (i.e. feature sufficient inertion). It is equally important to supply pure modelling techniques with well-developed scenario tools, the adequacy of which determines, to a large extent the efficiency of model application and the value of the obtained results. In scientific forecasting, as applied to human needs, it is necessary to account for the dynamics of changing material and spiritual needs, which in turn are considerably dependent on innovations, in the broadest sense of this term, and not only purely technological innovations.

A number of important methodological conclusions and recommendations have also been put forward during the discussions of specific projects and programmes accomplished at national and international levels.

For one, the Seminar participants showed a noticeable interest in the presentation of V.A. Kotelnikov, Vice-President of the USSR Academy of Sciences, entitled "The Comprehensive Programme of the USSR scientific and technological development and its social implications for 20 years" which, along with the general approach to the Programme elaboration, highlighted the most significant research and organizational problems. The VNIISI experts participating in the Programme development noted that the possibilities of obtaining relevant data and guidance on specific problems from the respective state agencies turns the very process of the Programme elaboration into a tool for identifying clearly the lines of scientific and technological progress to be followed in the light of the USSRs goals. Under socialism, said S.S. Shatalin, a corresponding member of the USSR Academy of Sciences, the elaboration of a programme of this kind is facilitated by the fact that it not only takes into account the social goals formulated in the official documents, but also the objective conditions under which scientific and technological progress must be directed towards their accomplishment. The development of the Programme requires an extensive interdisciplinary and interdepartmental cooperation, the enlistment of services and coordination of the activities of a huge number of scholars and experts. This of course is far from easy in practice. To overcome this objective difficulty, a number of specialized interdepartmental commissions were set up which, in the performance of their functions, use the knowledge and skills of experts in numerous areas of science and technology.

When discussing the "White Book on Research" prepared by a group of French scientists and representatives of state agencies, the speakers noted the importance of state control of basic research. The contribution of science to the solution of socioeconomic problems is dependent on a variety of factors, primarily on the socioeconomic structure, cultural traditions, level of scientific development, availability of an appropriate infrastructure and mechanisms of science organization and management, as well as on the specific country's goals and objectives, noted the participants.

Citing one of the statements in the "White Book" that the philosophy in contemporary science "goes its way in solitude", Dr I. Blauberg (USSR) noted that the position of philosophy is really unique, because—in order to be scientific—it cannot make its conclusions solely on the basis of scientific knowledge. This situation could be interpreted as a contradiction to the final part of the book, where a persistent idea is put forward consisting of the necessity to develop science in its wholeness and integrity, which is evidently impossible without the integrative role of philosophy. The key to solve this contradiction, pointed out Dr I. Blauberg, lies in a new phenomenon in contemporary science—the development of generalistic trends and interdisciplinary approaches pertaining to science as a whole, including all scientific subjects and branches of modern science, both natural and humanistic. Through these approaches specific scientific knowledge can be linked to, and integrated with, the philosophical reflections of science as a whole. Such approaches contribute to the synthesis of scientific knowledge and, as was demonstrated by systems methods, are necessary for facing large-scale complex problems challenging contemporary science.

Noticeable interest was shown in the positive and negative experience gained in the implementation of a number of international projects such as Interfutures, carried out by a group of scientists from OECD member-states, and the FAST project. In particular, those having been active in their development singled out the organizational difficulties arising from the activities of multinational research teams which reflected the contradictions and lack of agreed-upon positions of the governments concerned as to the issues under study. In addition, a regret was expressed that insufficient account had been taken of social and political factors, as well as of the interests of different social groups and various strata of society, thus considerably reducing the value of the obtained results.

One of the key directions of current scientific and technological progress is the automation of production, wide-scale employment of computers for technological system control, miniaturization of electronic equipment, and the growing introduction of microprocessors into production and non-production spheres. All this provides a basis for an ever-growing utilization of industrial robots.

The aforementioned processes are fraught with serious social consequences. Under respective social conditions the more extensive automation of production provides wide opportunities for a greater satisfaction of the material and spiritual needs of people, and the comprehensive development of personality. But if there is no adequate social orientation they threaten to produce a sharp reduction of jobs, unemployment and the estrangement of the individual from society. These issues were covered in the presentations of Dr A. King (IFIAS) and Prof. R. Ayres (USA).

By and large, broad discussion of the use of forecasting techniques in general, and of modelling in particular, as developed so far indicated that there are many methodological problems to be overcome. At the same time, as was also emphasized, these methods are needed not only for adequate investigation of contemporary problems but also in practical decision-making processes. Hence, further improvement of these techniques is imperative to make them more effective for practical implementation. In this respect it was unanimously agreed that a closer international cooperation in this field, including not only the mutual exchange of information and experience but also the organization of joint projects permitting full utilization of world scientific potential, is strongly needed. Only such joint efforts could contribute to the use of experience gained in various countries and scientific institutions, both national and international. The participants recommended strongly that an active role in promoting and developing this cooperation in the field of scientific forecasting related to the problems of human needs should be played by Unesco as a unique organization with an interdisciplinary mandate covering both problems of science and human development.

Dr O. Zakharova

Appendices

Appendix I
Final document

Unesco's 21st General Conference stressed the importance of the continuous review on trends in scientific research in relation to human needs and social goals. These activities have been conducted under the Unesco Programme on Research and Human Needs ("Contribution to the determination of research priorities linked with human needs and societal goals").

In order to fully implement the recommendations of the General Conference, as well as of those of several advisory panels* and to complement the already ongoing Unesco Programme it is recommended that a group of activities related to long-range scientific forecasting be initiated, in cooperation with national and international institutions. The symposium sponsored by Unesco, the USSR Commission for Unesco, the USSR State Committee for Science and Technology, the USSR Academy of Sciences, the All-Union Institute for Systems Studies and also with the cooperation of the International Institute for Applied Systems Analysis (IIASA), took place in Tbilisi, the capital of the Georgian Soviet Socialist Republic on 6-11 December 1981.

The general theme of the Symposium—"Scientific Forecasting and Human Needs: trends, methods and message"—has been prompted by the ever-increasing role science is playing in all aspects of global development. Consequently, long-range scientific forecasting will not only be necessary for monitoring the general dynamics of global evolution, but also will be of paramount importance in all the fields related to human development such as health, economics, education, etc. In particular, long-range scientific forecasting will be an indispensable tool for a better understanding of global problems and for trying to respond to the challenges, needs and threats which humanity will face in the future.

The participants of the Symposium have been guided in their discussions by the following basic considerations:

(1) the quest for a new economic order and the recognition of the cultural

*Including the Advisory Panel on Science, Technology and Society convened by the Director General in August-September 1981.

diversity of the world, opening a new phase of international cooperation which, among others, has the objective of improving the conditions of human life in developing countries and a marked shift in world development priorities;

(2) the paramount importance of the responsibility of the international scientific community for achieving a more humanistic reorientation of scientific research towards human needs and social goals;

(3) the necessity of curbing the arms race to make the ultimate goals of long-range scientific forecasting consistent with the establishment of a peaceful world;

(4) in long-range scientific forecasting the dynamics and evolution of human needs, as well as the fact that they are strongly related to local socioeconomic conditions, should be taken into account.

The discussion touched upon the following major aspects of scientific forecasting:

(1) comparison of methodologies in long-range scientific forecasting so far used in projects such as "Global 2000", "Livre blanc", "Interfutures", "FAST" and the "Soviet comprehensive programme for scientific and technological progress";

(2) methodological aspects, including systems approach, of long-range scientific forecasting in relation with human needs;

(3) priority areas for scientific forecasting related to human needs;

(4) a means for the practical implementation of forecasting results.

A summary of the discussions is attached in Note 1, and a publication is expected shortly, to be widely distributed to decision-makers as well as within the scientific community.

As the UN Agency with the multidisciplinary mandate of promoting scientific, cultural and educational activities, Unesco is uniquely placed for carrying on the activities in the field of long-range scientific forecasting, taking into account social and cultural implications of science and technology in global development.

To assist Unesco in the implementation of resolutions and recommendations of previous working groups and advisory bodies, several suggestions were made and a brief summary of those related to forecasting is attached in Note 2.

They led to the recommendation that Unesco, in order to complement the already ongoing Programme, initiates activities on long-range scientific forecasting related to human needs, emphasizing interdisciplinarity and a necessity for active international cooperation between the scientific institutions, both national and international, conducting research in this field. In order to implement these proposals it is recommended that:

(1) A working group be established to organize and coordinate these activities.

(2) A network of cooperating institutions, including United Nations agencies, national and international scientific institutions such as the International Council of Scientific Unions (ICSU), the International Federation of Institutes for Advanced Studies (IFIAS), the International Institute for Applied Systems Analysis (IIASA), the Institute for

Systems Studies (USSR), and national academies of science and research councils, be set up.

(3) Special attention be paid to the dissemination of the results of these activities through the communication media including press, films, television.

The participants of the meeting then expressed their warmest thanks to the Government of the Georgian Soviet Socialist Republic to the State Committee for Science and Technology and the Academy of Science of the Georgian Soviet Socialist Republic for their assistance and very kind hospitality. They went on to thank Unesco, the USSR State Committee for Science and Technology, the USSR Academy of Sciences, the Institute for Systems Studies and IIASA for sponsoring and organizing the symposium.

NOTE 1

As mentioned in the report, the discussion touched upon the following subjects:

(1) comparison of methodologies in long-range scientific forecasting so far used in projects such as "Global 2000", "Livre blanc", "Interfutures", "FAST" and the "Soviet comprehensive programme for scientific and technological progress;

(2) methodological aspects of long-range scientific forecasting in relation to human needs;

(3) priority areas for scientific forecasting related to human needs;

(4) a means for practical implementation of forecasting results.

The participants felt that special emphasis should be placed on those fields which have broad social implications, such as life sciences and material sciences. They acknowledged the special role played by natural resources and agreed on the necessity to focus on the future potential impact of resource deterioration, of long-term climatic change, and on development of new resources and on the new ways to use abundant resources in the satisfaction of human needs.

A major part of the discussion concentrated on methodological aspects of the long-term forecasting exercise. Full recognition was given of the usefulness of all major long-term modelling efforts for a better understanding of the global problematique and for increasing the general awareness of the public. These studies are to encourage the scientific community and young scientists in orienting their research efforts and studies, making them more consistent with social and humanistic objectives. The participants therefore underlined the necessity of further improving forecasting methods. Experience in forecasting methods shows gaps and weaknesses such as those listed below.

(1) Available methods are insufficient to take into account organizational and structural changes which occur within the time horizons used in the models. Hence projective power of models is weakened.

(2) Comprehensive, complex and large-scale models designed to answer many questions often end up giving no specific answers. Hence families of models, hierarchically structured, and each designed for a specific purpose, may have a greater value. The global problem forecasting tool

must be rather a system of interconnected models with variable structure, combining formalized and informal elements.

(3) The usefulness and efficiency of long-range forecasting depends to a great extent on availability and reliability of information. Hence information sources and data banks are of great importance.

(4) In order to increase the usefulness of such efforts, some participants felt that earlier involvement of policy-makers in the project may be especially helpful. More intensive dissemination of the project's results through academic and public media could enrich the educational efforts both at the university and general public level, thus contributing to a wider understanding of world problems and consequently providing a better ground for necessary policy changes.

NOTE 2

The following proposals were submitted to the working group appointed by the Symposium:

I. Conduct a retrospective analysis of former forecasts, especially those projecting technological trends, to identify the features of the most effective forecasting methods.

Develop techniques that could contribute towards anticipating the growth and saturation of human needs, material as well as intellectual and spiritual, in order to make peoples' expectations more consistent with carrying capacity and balanced global development.

Develop techniques that can use qualitative information on socioeconomic factors, such as scenarios, and explore procedures for using this qualitative information in combination with formal, mathematical models.

Conduct research on the consequences, for meeting human needs, of success or failure in the implementation of disarmament programmes. Develop methods that would allow trends in scientific research in long-range forecasting to be taken into account.

II. Forecast in the field of natural resources those which in the future are likely to be of reduced or increased utility.

Create an international programme for forecasting the development of ecological systems.

Develop a family of models that would be helpful to urban managers, especially in assisting the regulation of the "metabolic" processes of cities and in managing breakdowns.

Forecast the development and geographical distribution of research centres.

Integrate sectoralized models with appropriate scenarios of armament and disarmament processes.

Identify different categories of accidents generated by the intersection of the natural environment, technology, and society to define elements for adaptive behaviour and management.

III. Conduct a seminar of representatives of major scientific educational institutions which would use results of forecasting efforts to explore new modes of education that combine deep theoretical knowledge, broad engineering expertise, and ability to perceive and understand complex social problems.

IV. Develop better procedures for selecting the appropriate technique for forecasting a particular socioeconomic development. Create better awareness of limitations to the use of formal forecasting techniques. Work out improved methods for linking the results of forecasts with the administrative procedures actually necessary to improve the satisfaction of human needs.

Appendix II
Members of the Symposium Committees

INTERNATIONAL COMMITTEE

Academician J. GVISHIANI,
Deputy Chairman, USSR State Committee for Science and Technology,
Director of the Institute for Systems Studies
(USSR)

Dr H. HOGBE-NLEND,
President, Union mathematique africaine
(France)

Dr A. KADDOURA,
Assistant Director-General for Science,
Unesco

Dr A. KING,
President, International Federation of Institutes for Advanced Study
(IFIAS)

Dr A.M. M'BOW,
Director-General,
Unesco

Dr C. MALLMAN,
Fundacion Bariloche
(Argentina)

Dr F. MAYOR,
Universidad autonoma de Madrid
(Spain)

Dr A. PECCEI,
President, Club of Rome,
(Italy)

Dr S. RADHAKRISHNA,
Scientific Secretary, International Council of Scientific Unions Committee on Science and Technology in Developing Countries (ICSU/COSTED)
(India)

ORGANIZING COMMITTEE

Dr A. FORTI,
Division of Scientific Research and Higher Education, Unesco

Prof. S. KAPITZA,
USSR Academy of Sciences

Dr O. ZAKHAROVA,
Institute for Systems Studies
(USSR)

SOVIET ORGANIZING COMMITTEE

Chairman of the Soviet Organizing Committee
Acad. J. GVISHIANI,
Deputy Chairman, USSR State Committee for Science and Technology,
Director of the Institute for Systems Studies

Deputy Chairman
Dr S. KAPITZA,
Professor, USSR Academy of Sciences

Deputy Chairman
Mr. O. CHERKEZIA,
Deputy Chairman, Council of Ministers of the Georgian Soviet Socialist Republic

Dr A. BYKOV,
Deputy Chairman, USSR Committee for Systems Analysis,
Deputy Director, Institute for Systems Studies

Acad. I. GVERDTSITELI,
Vice-President, Academy of Sciences of the Georgian Soviet Socialist Republic

Mr Y. KASHLEV,
Deputy Chairman and Secretary-General,
USSR Commission for Unesco

Acad. V. KOTEL'NIKOV,
Vice-President, USSR Academy of Sciences

Mr G. KOTOV,
Director, Department of International Relations,
USSR State Committee for Science and Technology

Prof. B. MILNER,
Deputy Director, Institute for Systems Studies

Prof. S. MIKULINSKY,
Corresponding member of the USSR Academy of Science,
Director, Institute of the History of Science and Technology

Acad. Y. OVCHINNIKOV,
Vice-President, USSR Academy of Sciences

Dr S. TSUKANOV,
Director, Department of International Organizations,
USSR State Committee for Science and Technology

Acad. E. VELIKHOV,
Vice-President,
USSR Academy of Sciences

Executive Secretary
Dr O. ZAKHAROVA,
Institute for Systems Studies

Index